T0134011

PROCEEDINGS OF THE 30TH INTERNATIONAL GEOLOGICAL CONGRESS
VOLUME 25

MATHEMATICAL GEOLOGY AND GEOINFORMATICS

PROCEEDINGS OF THE 30TH INTERNATIONAL GEOLOGICAL CONGRESS

PROCEEDINGS OF THE
30TH INTERNATIONAL GEOLOGICAL CONGRESS

BEIJING, CHINA, 4 - 14 AUGUST 1996

VOLUME 25

MATHEMATICAL GEOLOGY AND GEOINFORMATICS

EDITORS:
ZHAO PENGDA
CHINA UNIVERSITY OF GEOSCIENCES, WUHAN, CHINA
F.P. AGTERBERG
GEOLOGICAL SURVEY OF CANADA, OTTAWA, CANADA
JIANG ZUOQIN
MINISTRY OF GEOLOGY AND MINERAL RESOURCES, BEIJING, CHINA

UTRECHT, THE NETHERLANDS, 1997

VSP BV
P.O. Box 346
3700 AH Zeist
The Netherlands

© VSP BV 1997

First published in 1997

ISBN 90-6764-268-1

Printed in The Netherlands by Ridderprint bv, Ridderkerk.

CONTENTS

Proc. 30th Int'l Geol. Congr., Vol.25, pp. 1-12
Zhao Peng-Da et al (Eds)
© VSP 1997

The Development of Geological Hazards Map and Image Analytical System (GHMIAS) and Its Application in Land Subsidence Analysis in Tianjin, China

SHI JIANSHENG,   ZHANG FENGBIN,   CHENG YANPEI

Institute of Hydrogeology & Engineering Geology, M.G.M.R.,
Zhengding, Hebei, 050803 P.R.China

Abstract   The distribution, occurring and developing of geological hazards are spatial, the influencing factors of which have their own features and complexity. The GHMIAS developed by the author is a kind of applied GIS with typical geological hazards as its key objects, and with management and analysis of spatial information as its main functions. This system has unique spatial analyzing model expansion. It combines the vector and raster format, permitting mutual transform of various data formats. With its manifold mapping functions and high quality map output, it can produce thematic maps quickly. Lots of analyses have been made on hazards distribution, occurring mechanisms, influencing factors as well as developing trends of regional geological hazards in Beijing-Tianjin-Tangshan area, and valuable results of land subsidence process and developing trend in Tianjin city have been obtained. These applications show that GHMIAS can play an important auxiliary role in solving geological hazards distribution, occurring and developing regularities, prediction of their evolution.

Key words: GIS,   spatial analysis,   geological hazards,   prediction

INTRODUCTION

Geological hazards are serious natural disasters concerning with human existence. Their distribution, occurring and evolution are spatial, and the influencing factors have their own features and complexity. Graphic systems suitable for thematic spatial information management and analysis of geological hazards are scarce so far. Special research on geological hazards map and image auxiliary analytic system were carried out during the "Eighth Five-year Plan" period. The purpose of the research was to develop a geological hazards map and image analytic system (i.e. GHMIAS) with map-image input, storage, process, display, analysis and output functions on the basis of incorporating the strong points of various GIS systems and other map analytic systems.

Geological hazards are always related with certain space range. The influential space of geological hazards we observed is the result of many complex factors, which comprises comprehensive natural and artificial relations. GIS provides us the probability of analyzing the relations and interaction results of various factors[2,3,4]. GHMIAS is an applied thematic system in the field of spatial information management and analysis of geologic hazards, and can be used in geological hazards prediction and control as well as policy making.

In the light of the characteristics of geological hazards, this system provides spatial evaluation and prediction, graphic presentation, thematic maps production and process in addition to the conventional functions, such as data collection, storage, analysis and output. Its graphic data structure and database system design reflect the professional information features in order to utilize the information more efficiently and fully.

In the application analysis, we attempted to provide a scientific basis for geological hazards evaluation and prediction, and provide basic parameters for other precise mathematical models from the viewpoint of spatial statistics, overlay and model prediction to meet the requirements of geological hazards prediction, prevention and control[1].

1.   DEVELOPMENT OF GHMIAS

1.1   Supporting environment

Hardware: PC 386 or above (Pentium 586 is recommended); 4M memory at least (8M is recommended); 40M bytes free in hard disk at least; any input and output devices supported by Windows.

Software: Dos 5.0 or above; Chinese Windows 3.1 or above, or Chinese Win95.

1.2   Composition and main features

1.2.1 Composition

Composition of GHMIAS is shown as below(fig.1):

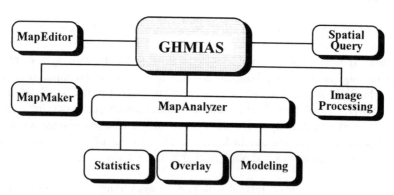

Fig.1   **Sketch map showing the components of GHMIAS**

1.2.2 Main features

GHMIAS, an analytic GIS with geological hazards as its main objects, has such features:

(1)   It has conventional functions of GIS, such as map input, editing and management, inquiry display, analysis processing and output;

(2) GHMIAS has many newly designed analysis models, such as image overlay, spatial Gray system modeling, which has enriched the types of spatial analytic models and enhanced the analysis capability;

(3) GHMIAS has advanced data structure, combining vector, grid and graph user's objects of Windows together, supporting mutual transform of various data formats, and can share data with ARC/INFO, INRISI, SPACEMAN and other GIS systems home and abroad;

(4) GHMIAS has visual interface and is easy to operate by adopting the pattern of Microcomputer plus Chinese Windows platform, which is in accord with the trend of software development and the requirement of popularization.

1.3 Layer structure data model

Data model and data structure are the core of GIS, and the crux of realizing the functions completely and flexibly. The objects of GHMIAS are the information that has certain space features and complex attribution relations. The purpose of the designing of data model and data structure was to set up a professional database, and to lay a foundation for specialized GIS.

(1) Project

Project is the most top object of information management set up for specific purpose in certain information field, and also the database collection concerning specific field and purpose. Thus a project manages and controls more than one database.

(2) Database

Database is the assemblage of various information files with different storage formats. A database is composed of more than one file.

(3) File

File is the basic unit when computer operation system manages user's information. In GHMIAS's data structure, data file is an information unit composed of more than one layer.

(4) Layer

Any graphic file is composed of information of various attributions. In order to distinguish different attributions and process every attribution data respectively, layers are introduced to manage the different attribution data.

(5) Element

Element is the smallest unit of graphic information. GIS system developed in Windows environment can form the information record system that combines vector, grid and Windows standard elements. User's information usually is represented by point, line and polygon, having the characters of vector; space image, photo and other scan images are represented by grid; regular figures provided by Windows system, rectangular and eclipse for instance, can be used as signs, notes and map making. The coexistence of three kinds of elements enhances the operation performance, graphic presentation effect and makes the output more easily.

Thus we set up the information chain structure: project → database → file → layer → element . By such a chain structure, users can easily set up complex information structure and form complete information network under the guidance of the system beginning with creating a project; and the system can automatically adjust when users update or edit the information of any layer.

1.4   System module and function

GHMIAS has formed the preliminary framework of functional map and image analysis system. GHMIAS is composed of Vec_map Editor module, Raster Analyzer module, Map Maker module, Map Browser module, Dbase Editor module and Map Helper module, including image overlay analysis model, image sequential Gray prediction model and many other analysis models as well as basic GIS functions. The established map and image database of geological hazards prediction and prevention in Beijing-Tianjin-Tangshan area comprised hundreds of layers concerning basic topography, social economy, present geological hazards condition, space distribution of influencing factors, prediction and evaluation. The system can also produce high quality maps.

1.4.1 Functions of Vec_Map Editor

Multiple file can be opened at the same time to process maps. The main functions include:

> FILE: New, Open, Close, Save, Import, Export, Digitize, Print, Exit;
> EDIT: Select, Copy, Delete, Move, Rotate, Deform, Attribution, Scale, Change File Head;
> VIEW: Fit to Window, Zoom In, Zoom Out, Display Control;
> DRAW: Text, Point, Line, Polygon, Rectangular, Square, Circle, Eclipse;
> SETTINGS: Paper Size, Layer, Ruler, Text Property, Line Property, Fill Property, Point Types, Color Panel;
> HELP: Index, Subject Help, Glossary, About This Module.

1.4.2 Functions of Raster Analyzer

> FILE: New, Open, Close, Scan, Attribution Relation, Print;
> EDIT: Modify, Copy, Delete;
> DRAW: Same with Vec_Map Editor Module, but save as grid format;
> PERFORM: Transform between vector and grid, Attribution Inquiry, Modify File Head, Get Attribution From ···. Resort, Zoom In, Zoom out, Flip, Makeup.

Windowing, Filter;

STATISTICS: Histogram, Crosstabulation, Regression, Autocorrelation, Trend Analysis, Random Map Generating;

GRAPHIC ALGEBRA: Overlay, Constant Calculation, Area Calculation, Perimeter Calculation;

SPATIAL MODEL: Gray System Modeling, Distance Analysis, Expenditure Surface, Optimum Route, Demand and Supply Analysis, Sorting, Surface Analysis, Visible Analysis, Watersheds Analysis;

PROCESS: Recognition, Sorting, Standardization, False-Color Composite, Strip Removal, Filter, Main Composition Analysis, Fuzzy Matrix Analysis.

1.4.3 Map Maker

GHMIAS supports the ornament of vector and grid maps, pooled output, and many output devices, from stylus printer and laser printer to large pen plotter and large color ink printer.

1.4.4 Map Browser

This module is a sorting and inquiry system of geological hazards map and image database of Beijing-Tianjin-Tangshan area. Just click on any point of the map, you can access the related map and image database.

After completion of the revision of the geological hazards map and image database of Beijing-Tianjin-Tangshan area, plus the support of the module, the system will be beneficial to the prediction and prevention of geological hazards of this area.

1.4.5 Map Helper

This module is a convenient tool for users to learn how to use the system. Users can get help from this module and the Helps of other modules whether you are familiar with the system or not, getting information of data structure and file structure.

2. APPLICATION OF GHMIAS IN LAND SUBSIDENCE ANALYSIS OF TIANJIN CITY

2.1 Time-space Statistics and Evaluation of Water Level Changes

Previous study showed that the cause of land subsidence accelerating in Tianjin was the continuous ground water level drop by overpumping. Since measures were taken in 1980s, the land subsidence was under control[5,6,7,8].

As ground water level change is the direct inducing factor of land subsidence, lots of analyses have been made, most of which were direct statistic analysis of scattered observation data. The actual information is continuous in time and space, but what we collected from the nature is noncontinuous and disperse. GHMIAS can simulate a continuous space through dippers interpolation and isarithmic interpolation according to the disperse information collected. So the result is closer to the natural condition, and more valuable to the policy-making.

Our water level analysis in Tianjin is based on the observation data of the second and third aquifers in 1980, 1985, 1988. The basic analysis process is shown in Fig.2.

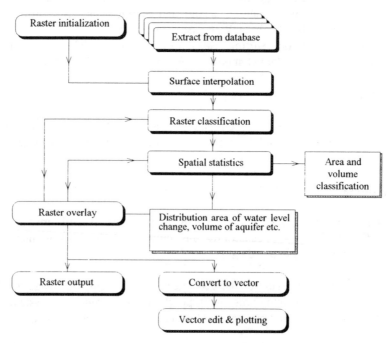

Fig. 2   The basic processes of spatial subsidence analysis in Tianjin

The statistic result (Tab.1) shows that the decreasing trend of ground water level in both second and third aquifers was relieving in 1980s. As for the second aquifer, whether the absolute values (maximum, minimum and average value) or the ratio of the area that water level was increasing and the area that water level was decreasing or the moisture-laden volume all were developing toward water level increasing.

**Table 1. The feature values of groundwater level in 80s
in urban and suburban districts of Tianjin City**

aquifer	period	max. (mm)	min. (mm)[*]	mean (mm)	standard deviation
II	80—85	21.33	—34.67	—3.67	6.29
II	85—88	49.0	—14.4	6.3	9.62
II	80—88	31.47	—20.0	2.63	10.06
III	80—85	31.5	—37.0	—5.57	10.25
III	85—88	41.0	—21.0	4.79	9.65
III	80—88	33.1	—36.0	—0.77	11.38

[*] the value in negative sign represents water level decrease in the specified period.

The total increase of ground water level surpassed the total decrease at the end of 1980s. The intensity of water level decrease in the third aquifer was diminishing, but on the whole the total decrease of water level surpassed the total increase, which showed the water exploitation control measures in this aquifer was not very strong.

2.2   Spatial Analysis and Evaluation of Surface Subsidence

The same principle as in water level analysis was employed to analyze the surface subsidence according to the observation data in 1985, 1988, 1990, 1992. The result (Fig.3, Fig.4 and Tab.2) showed: the effect of control measures was obvious around the year of 1988 with the compacted volume of soil mass decreased greatly; after 1988, surface subsidence kept relatively stable, the accelerating rate was the least in 1990. Around 1992, with the depletion of the function of water level control measure of the second aquifer, and new subsidence factors were not under control properly, subsidence began to rebound.

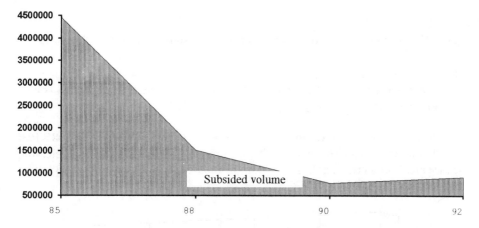

Fig.3   Volumetric changes of subsidence in Tianjin from 1985 to 1992

Table 2.   Statistic values of subsidence from 1985 to 1992

year	min.*	max.	mean	standard deviation
1985	14.0	146.0	80.82	18.121
1988	-16.0	113.0	27.163	21.211
1990	-5.0	53.0	14.09	8.4424
1992	-3.0	62.0	16.55	12.0143

*
  the value in negative sign represents surface rebound during the specified year.

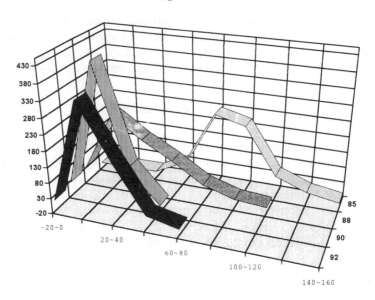

Fig.4   Distribution of classified spatial statistics of subsidence from 1985 to 1992

The distribution of rebound value showed the same result: the rebound value around the year of 1988 was the biggest, after that the value began to decrease; the rebound volume in 1992 further reduced a lot comparing with that in 1990.

The retarded subsidence area was diminishing since 1988, from 549.34 km^2 in 1985-1988 to 418.00 km^2 in 1988-1990, and to 283.11 km^2 in 1990-1992; while accelerating subsidence area was increasing, from 2.75 km^2 in 1985-1988 to 134.09 km^2 in 1988-1990, and to 268.98 km^2 in 1990-1992 (Fig.5).

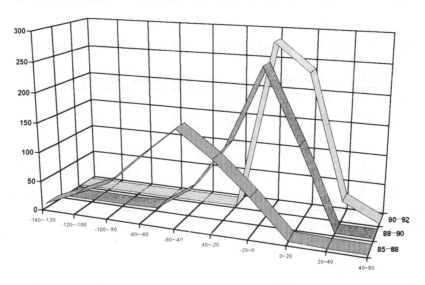

Fig.5 Distribution of the classified spatial statistics of subsidence vary in period

Table 3.   Statistic values of subsidence changes in different period

period	min*	max.	mean	standard deviation
1988—1985	-138.47	21.19	-53.65	27.8823
1990—1988	-77.94	20.49	-13.08	16.2268
1992—1990	-14.95	43.00	2.46	9.6924

* the value in negative sign represents decrease in value of subsidence in the specified period.

The characteristic values of subsidence change in different periods (Tab.3) also revealed that the rebound was decreasing while subsidence was increasing since 1988. The average subsidence value of Tianjin area (including urban area and outskirts) raised 2.46 mm in 1992 comparing with that in 1990.

Subsidence center and subsidence value of different periods varied from place to place. Subsidence in 1988 weakened than that in 1985. Gongrenxincun, Fangxinzhuang and the perfumery to the east of downtown were the areas that abated most significantly. The condition in 1990 did not change much comparing with that in 1988. Of which, subsidence in most places of downtown and Liming village-Huantuo area in the north-east increased slightly while in other places decreased continually. In 1992, subsidence velocity kept stable in most places, but decreased slightly in downtown, north-east and north-west areas, and increased obviously in the south-west, village Huazhuangzi.

2.3   Gray Spatial Model Prediction

A demonstration prediction analysis was made by GHMIAS with its Raster-based Gray Spatial Prediction Model according to the subsidence observation data of 1985, 1988, 1990 and 1992.

SGM is actually a gray system model set which is based on time-series raster data sequence. The basic analytical steps are as follows:
If we have a time-series raster set: $\{R(t)\}$, $t=1,2,...,n$
To the certain location, the original data sequence is: $\{X^{(0)}(t)\}$, $t=1,2,...,n$
After doing simple add up operation, we get a accumulated data sequence:

$$\{X^{(1)}(t)\}, \ t=1,2,...,n. \quad \text{In which:} \quad X^{(1)}(t) = \sum_{k=1}^{t} X^{(0)}(k)$$

According to professor Deng[9], at the time point of (t+1) the prediction value follows this formula:

$$\hat{X}{}^{(1)}(t+1) = (X^{(0)}(1) - \frac{\mu}{\alpha})e^{-at} + \frac{\mu}{\alpha}$$

In which, μ and a are extracted from differential formula:

$$\frac{dx^{(1)}}{dt} + ax^{(1)} = \mu$$

The real prediction value should be backed from the accumulated sequence:

$$\overset{\wedge (0)}{X}(t+1) = \overset{\wedge (1)}{X}(t+1) - \overset{\wedge (1)}{X}(t)$$

To judge the model precision, we use following two parameters:

$$① \quad C_m = \frac{1}{cols \times rows} \sum_{i=1}^{cols} \sum_{j=1}^{rows} C_{ij}$$

$$C = \frac{S_2}{S_1} = \frac{\sqrt{\frac{1}{N}\sum_k (e(k) - \bar{e})^2}}{\sqrt{\frac{1}{N}\sum_k (X^{(0)}(k) - \bar{X}^{(0)})^2}}$$

In which:

$$\bar{X}^{(0)} = \frac{1}{N}\sum_k X^{(0)}(k), \qquad \bar{e} = \frac{1}{N}\sum_k e(k)$$

$$② \quad P_m = \frac{1}{cols \times rows} \sum_{i=1}^{cols} \sum_{j=1}^{rows} P_{ij}$$

In which: $p = P\{|e(k)-e| < 0.6745 \ S_1\}$

The best prediction is the one with p_m greater than 0.95 and C_m less than 0.35. The worst prediction is the one with p_m less than 0.70 and C_m greater than 0.65.

A.   Gray Prediction on Surface Subsidence of 1994

In the light of the spatial statistics stated above, gray predictions on some characteristic values were made. Theories and technology concerned please refer to calculation principle of the Map Analyzer module. The result is shown in Tab.4.

Table 4. Characteristic values of subsidence predicted in 1994 and 1996

predicted items	data sequence			1994a prediction	1996a prediction	model test			
	year	original value	calculated value			C	P	S	precision
min. (mm)	88	-16.0	-16.0	-1.8	-1.1	0.006	1	0.682	GOOD
	90	-5.0	-4.92						
	92	-3.0	-2.98						
max. (mm)	88	113.0	113.0	72.32	84.6	0.0018	1	0.734	GOOD
	90	53.0	52.88						
	92	62.0	61.84						
mean (mm)	88	27.16	27.16	19.38	22.76	0.0024	1	0.734	GOOD
	90	14.09	14.06						
	92	16.55	16.51						
decreased area (km²)	88-85	549.3	549.3	191.63		0.016	1	0.69	GOOD
	90-88	418.0	413.7						
	92-90	283.1	281.5						
increased area (km²)	88-85	2.75	2.75	484.39		0.026	1	0.766	GOOD
	90-88	134.1	127.0						
	92-90	268.98	248.1						

From the result, surface subsidence volume of Tianjin city is 5,409,538 m³ in 1994, predicted rebound volume is 6,757m³; the minimum subsidence value is -1.8 mm(i.e. rebound value is 1.8), the maximum 72.32 mm, the average 19.38 mm; the area that subsidence velocity would increase is 484.39 km², the area that subsidence velocity would decrease is 191.63 km². The result shows that subsidence will further develop if strong subsidence-control measures are not taken. Because the time sequences of data are rather short, the correction capacity of the model is limited. Although the error bound is not satisfying, the simulating results show same trend with the actual data.

B. Distribution Analysis and Prediction of Subsidence Trend in 1994 and 1996
The analysis procedure is as follows:

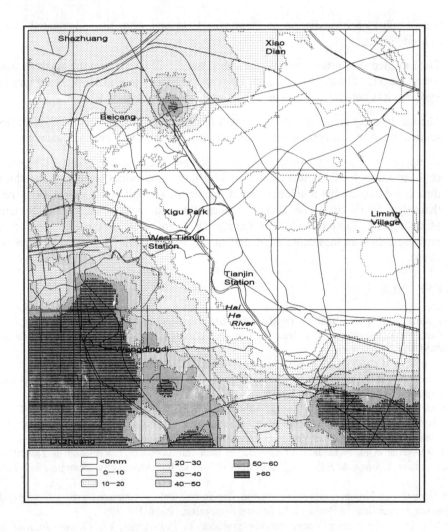

**Fig.6 Prediction map of subsidence in Tianjin in 1996
by the means of Spatial Gray System Modeling**

(1) Acquire observation data of subsidence of 1985, 1988, 1990 and 1992;

(2) Set up the characteristic surface of subsidence using the surface function of disperse data interpolation;

(3) Predict the subsidence in 1994 and 1996 according to four characteristic surfaces using Spatial Gray System Model (SGM)(Fig.6);

(4) Analyze the space of residual error to determine the fiducial interval and evaluate the precision and reliability of model prediction;

(5) Carry out statistic analysis, classified picking-up and calculation of area and volume through the grid map analysis function of GHMIAS.

The overall subsidence prediction of Tianjin in 1994 and 1996 through Spatial Gray System Modeling is:

(1) The subsidence in urban area will be stable, but it will increase in the outskirts to a certain degree. The average value of subsidence is 19.77 mm in 1994, 24.09 mm in 1996. So it is necessary to enhance the subsidence-control measures.

(2) The accelerating center of subsidence will further move toward Huazhuangzi in the south-west. And it is probable that the subsidence in the south-east will increase.

(3) Compared with the gray prediction on subsidence characteristic values, the result obtained through SGM model reveals enlarged extreme values because of the short time sequence, i.e. the absolute value of maximum and minimum values are larger than they really are. The average values obtained from Spatial Gray System Modeling (19.77 mm in 1994, 24.09 mm in 1996) are similar to those obtained from gray prediction on subsidence characteristic values (19.38 mm in 1994, 22.76 mm in 1996).

REFENRENCE

1. Shi Jiansheng, Spatial analysis on the relationship between active fault and Karst surface collapse in Tangshan city of China. *The Chinese Journal of geological hazard and Control*, **P87-91**,Vol.7 No.3, 1996.
2. Wu Lun et al, A Guide Book to GIS, Beijing university Press, **P1-5**, 1994.
3. J. T. coppock et al, Methods of spatial analysis in GIS, International Journal of Geographical Information Systems, Vol. 4 No. 3 1990.
4. Michael F. Goodchild, Karen K. Kemp, Introduction to GIS, NCGIA Core Curriculum, 1991.
5. Hu Huimin et al. The development of ground subsidence in main cities of North China, The Chinese journal of Geological Hazard and Control, **P1-9** Vol.2 No.4, 1991.
6. Cui Xiaodong et al. A mathematical model of land subsidence in Tianjin and its numerical computation, *Geological Hazards*, Proceedings of Beijing International Symposium, P310-312, 1991
7. Jin Dongxi, Niu Xiujun, Land subsidence in Tianjin city and the countermeasures. *Geological Hazards*, Proceedings of Beijing International Symposium, **P310-318**, 1991.
8. Li Minglang, The main geo-environmental problems in Tianjin city, *The Chinese Journal of geological Hazard and Control*, Vol.5 No.3, 1994.
9. Deng Julong, *Gray Control System*. Central China university of technology press, 1985.

Proc. 30th Int'l Geol. Congr., Vol.25, pp. 13-22
Zhao Peng-Da *et al* (Eds)
© VSP 1997

Quantitative Appraisal for Mineral Resources with Integrated Information

WANG SHICHENG[1], YU XIANCHUAN[1], CHEN MING[2]

1 Institute of Geomathematics, Changchun University of Earth Sciences, Changchun, Jilin, 130026, People's Republic of China

2 Geochemistry department, Changchun University of Earth Sciences, Changchun, Jilin, 130026, People's Republic of China

Abstract

The theory of Quantitative Appraisal for Mineral Resources with Integrated Information (QAMRII) includes three parts. The first is the theory of metallogenic prognosis. The second is the theory of statistical prognosis of resource occurrences. The third is the theory of quantitative prediction of mineral resources. QAMRII regards the geological hypothesis as geological prerequisite, the geological bodies and the possible mineral resources bodies as the basic prognosis units in prediction. Different kinds of geo – data are generally not unanimous. QAMRII studies the interrelation of geological, geochemical, geophysical, and remote sensing information for the prognosis of mineral resources, including larger and superlarge deposits and blind deposits, of known and unknown types. A wide range of methods to transform, associate, and synthesize the different kinds of information, to delineate the geological and mineral resource bodies, to select the model units, to choose the mathematical models, to determine the necessary variables are studied in QAMRII.

Keywords: mineral resources, information, prognosis, QAMRII, economic geology, theory of similarity analogy, theory of anomaly – seeking

INTRODUCTION

Geological, geochemical, geophysical, and remote sensing data are of inevitable multi – solution. Application of comprehensive information is the tendency of geological study. The concept of Geomathematics should be changed. We considered Geomathematics as a subject to study the interrelationships between the different kinds of geological information and the relationship between information and mineral resources bodies in 4 – dimensional space (X, Y, Z, T) by transforming, interpreting and synthesizing the geological, geochemical, geophysical, and remote sensing data in mathematical thoughts, methods, and models(4 – 5). obviously, this is an important principle in geological studies.

The theory of QAMRII includes three parts: 1)the theory of metallogenic prognosis,

2)the theory of qualitative prognosis of the occurrences of mineral resources, and 3)the theory of quantitative prediction of mineral resources(4 - 6). Under the guidance of the theory of metallogenic series [2 - 3, 5 - 7), the different kinds of geo - data are interpreted strictly depending on their own basic principles. Regional gravity and magnetic data are used to ascertain the tectonic frame of basement that controlds the distribution of overlying strata. The vertical derivatives of the second order are applied to delineate occurrence of intrusive bodies, and their size and shape. Remote sensing data are employed to find linear and annular structures. Regional geochemical data are used to define the favorite areas. Geological maps are regarded as the premise. All kinds of information mentioned above are appraised, filtered. The significant variables are retained and used to determine the location of mineral resources, the possible number of deposits or ore bodies, and the amount of mineral resources, etc.

SUMMARY OF THE PRINCIPLE OF QAMRII

Genesis of ore deposits has its rules. Hence, the hypothesis of metallogenesis must be regarded as a priori prerequisite. Though these theories themselves are not perfect, they came from huge number of geological investigations and at least partially reflect the law of processes in the geologic time.

Grid units were commonly used in mineral resources prognosis before the theory of QAMRII was brought into operation. We incline to use the geological units defined by geological bodies. A geological unit is an area in which the geological and metallogenic conditions are relatively accordant. Statistically, mineral resource bodies are special kinds of geological bodies, and an ore deposit is an aggregate of special rocks. Prognosis based on geological units can wholly reflect the complete characteristics of mineral resources.

One of the obstacles in the interpretation of geo - information is that one kind of geo - data can only represent one aspect of a natural geological body. Different kinds of geo - data are usually ambiguous from one to another. Most kinds of geo - data are of multi - solution inevitably. It is the key to "conjoin" the unanimous data and explain it properly. QAMRII studies the interrelationship between different types of information. A series of methods is used to explain, transform, and synthesize the information of field investigation, geochemistry, geophysics, and of remote sensing to predict the possible deposits of known and/or unknown types. The principle of similarity - analogy is used to explore the deposits of known types while the principle of anomaly - seeking is applied to detect the deposits of unknown types. Under the guidance of the theory of metallogenic series, QAMRII has also been employed to predict large and superlarge mineral deposits. As the result of application of QAMRII, a good number of deposits have been found and apparent economic benefits have been produced in China.

The Reliability of the results of QAMRII depends on the level of information drawn from the investigated geological, geochemical, geophysical, and remote sensing data. The following points must be stressed(1 - 3):

(1). The results of geological investigation in the field must be tested at first. Generally, field investigations are not as precise as mathematics, physics and chemistry. They are limited by the complexity of geological phenomena and geologist's experiences. Different geologists would draw different conclusion about the same geological object.

Geochemical, geophysical, and remote sensing information can be used to proved out the reliability of geological maps. The difference between above – mentioned kinds of information is unavoidable, only integrated information can interpret them reasonably.

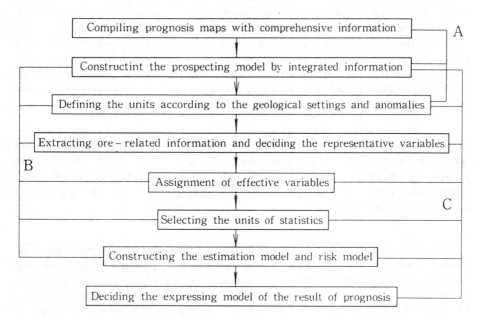

Fig. 1 The flow chart of QAMRII. Symbols : A, metallogenic prognosis; B, qualitative prognosis of the locations of mineral resources; C, quantitative prediction of mineral resources

(2)According to our experience, different types of geological objects have similar physical and/or chemical properties while the similar physical and/or chemical properties represent different types of geological bodies. Geophysical and geochemical fields and anomalies are caused by geological and resource bodies, but the investigated values cannot be used directly. Only the comprehensive interpretation of integrated information according to the geological and resource units can reveal the essence of geological bodies.

(3). The unbalance of economic development results in the unbalance of geological study in different area. Mineral resources prognoses permanently meet the difficulty of unbalanced information that cause the difficulty in application of the theory of similarity – analogy. However, lower level of study does not mean nonexistence of deposits. To predict the mineral resources in the undeveloped area according to the models established in the developed area, the relationship between different sets of information of geological and resource bodies must be studied.

(4). The concentration of element and mineralization of inheritable. Studies in various areas have shown that the great content of an element in the ancient basement would cause the younger igneous rocks and sedimentary strata to be rich in this element. This element would participate in the later processes of mineralization, too. The metallogenic series related to metamorphic system occurred first, and then the other series. If the material source was not the mantle, the later was developed on the former. In south China, the copper deposits in sandstone system distribute along the edge of palaeobasement. The metamorphic basement in Jiangxi Province was rich in W element. The chemical compositions of the younger strata are rich in W element, too. The subsequent tectonism, magmatism and metallogenesis reformed the stratum's system and finally resulted in deposits, including the Xihuashan superlarge tungsten deposit. through hydrothermal activities.

(5). The generality and individuality of ordinary geological bodies and resources bodies are the basis of comparison and division of them. The rules of comparison and division are similarity – analogy and anomaly – seeking principles. What we emphasize is that both of the principles of similarity – analogy and anomaly – seeking must be applied in the same time.

(6). Ordinary geological bodies and mineral resource bodies are corresponding to the geological settings and anomalies related to mineralization respectively. Geological anomalies(8 – 9) are the sources of geophysical. geochemical. and remote sensing anomalies. The departed geological investigations usually stressed the former and neglect the later. Mineral resources bodies must be geological anomalies. Geochemical. geophysical, and remote sensing anomalous characters are their profiles. The theory of QAMRII stresses the importance to comprehensively evaluate the geological anomalies against the geological settings, and its flow of actualization and the corresponding methods can be summarized as Fig. 1 and Tab. 1.

METALLOGENIC CONDITIONS RESEARCH

The intent of compiling prognosis maps is to confirm the geological bodies in 4 – dimensional space. Geological maps reflect the combination of geological and resources bodies of various properties and scales formed in geological periods. The information drawn from geophysical. geochemical, and remote sensing data also partially reflects the geological bodies. The causation between the measured data and geological bodies will be clear after the statistical comparison of geological maps with the geophysical. geochemical, and remote sensing maps. The methodology for comprehensive interpretation includes five aspects:

(1). Deduce the buried Precambrian metamorphic bodies of basement from the gravity data. Generally, the density of metamorphic Precambrian is greater than ordinary sedimentary, volcanic, and intrusive rocks, if their chemical compositions are the same The linear beddings of magnetic anomalies represent the fundamental distribution of the ancient basement. Positive magnetic anomalies or strongly jumping magnetic fields escorted by positive gravity anomalies surely indicate the existence of orthometamorphite, and negative magnetic fields indicate the parametamorphite. In the mainland of China, the basement is mainly in EW direction. This direction is one of the directions of mineralization.

(2). Sketch the regional structure frame in the viewpoint of geological evolution. Gravity, magnetic, and remote sensing data are used to outline the structural skeleton of studied areas. The linear characteristic lines drawn from the horizontal derivative of first order and the annular patterns drawn from the vertical derivative of second order of gravity and magnetic data are synthesized with the linear and annular structures drawn from remote sensing data respectively. Under the premise of geological hypothesis, the order of generation of regional structures can be figure out according to the gravity and magnetic index beds and their generating order. Anomalous orders of structure are usually related to ore bodies, deposits, orefields and mineralization domains.

(3). The structural frame of basement controlled that of the capping bed. Although both of the overlaying volcanic rock and sedimentary rock inherit the chemical composition from the ancient basement, they have different characters. The pure sedimentary rocks have negative magnetic field while sediment volcanic rocks have strongly jumping magnetic field.

(4). The tectonic frame of basement controlled the structures as passageways for intrusive rocks. So did the gradient zones of Moho surface. Different intrusive rocks have their own magnetic characteristics. Especially, the intermediate and/or intermediate – acid intrusive rocks can be divided into non – inherited rocks and inherited rocks[7, 8]. The non – inhabited rocks have simple magnetic field while the other partially inherit the magnetic characters of basement. Inherited rocks are the result of multi – magmatism. Their shapes and attitudes can be decided on the vertical derivative of second order of magnetic data. Large gold deposits of alteration are controlled by the inherited intrusive rocks.

(5). Studying the river system with remote sensing data and synthesizing the different kinds of information in the viewpoint of neotectonism. Practically, the distributions of linear and annular drainage basins are harmonious with the linear imagines and the linear and annular characteristic lines drawn from gravity and magnetic data. They also reflect the skeleton of structural system and control the distribution of primary and secondary geochemical holes, the sedimentary and heavy mineral anomalies. The occurrence of their intersection, especially the intersection of linear and annular characteristic lines, indicates the favorite minerogenetic area[6 – 7].

Wang Shicheng et al.

Tab. 1 The methods in QAMRII

Items	Mathematical models or methods
Compiling the prognosis maps	Horizontal and vertical derivatives of gravity and magnetic data. Upward (downward) continuance analysis. Moho surface calculation. Multivariate statistical analysis
Metallogenic model with comprehensive information	Transformation of potential field. Multivariate statistical analysis. Fuzzy analysis Robust statistical analysis. Geostatistical methods. Fractal and entropy analysis
Variables' selection	Multivariate statistical analysis. Fuzzy and Entropy analysis. Quantitative theories
Assignment values the selected variables	Multivariate statistical analysis. Fuzzy and rank analysis. Quantitative theories. Contingence table analysis
Units' selection	Multivariate statistical analysis. Quantitative theories. Optimization partitioning. Multidimensional scaling
Occurrences of mineral resources prediction	Fuzzy analysis. Geometry probability. Characteristics analysis. Multivariate statistical analysis Logistic regression. Quantitative theories. Gray system analysis
QAMRII	Neural networks. Logical information methods. Weighted characteristics analysis. Quantitative theory. Probability regression. Geostatistics. Monte – Carlo simulation Poisson regression

ESTABLISHING THE COMPREHENSIVE PROSPECTING MODELS

The basis of prospecting model is the metallogenic models of ore bodies, ore deposits, and orefields. Synthesizing information of geology, geochemistry, geophysics, and remote sensing, selecting the substantial information, eliciting the key factors of mineralization, and finding the signs of ore deposits are the basic steps of modeling program.

A geological prospecting model is a statistical model of one sort of deposits based on comprehensive information. It depicts the relationship between geological processes and mineralizations, and is applied to predict deposits of known and unknown kinds.

Under the guidance of the theory of metallogenic series [2 – 3], the models of metallogenic series can be implemented through the study of the series of deposits in different areas with different geological setting. The comprehensive prospecting models are based on geological prospecting models and models of metallogenic series.

SELECTION OF STATISTICAL UNITS

The selection of statistical units depends on the predicting scale and the relationship between geological bodies and mineral resources bodies. The delimitation of units must be agreeable to the aim of prognosis. There are different sizes of geological bodies, from as large as a orefield to as small as an orebody. If the aim of prognosis is to estimate the total amount of mineral resources for government's policies, units of the size of orefields will be selected. If the aim is to find orebodies for a mining company, units of the size of orebodies will be used. For a part of geological bodies are buried, the corresponding units are 'blind', too. The boundaries of the blind unit can be delineated only by integrated information.

There are four rules for statistical units division: (1) the geological condition in each unit must be uniform relatively, one unit is from one gross; (2) different kinds of units have notable variances; (3)the amount of units must satisfy the demands of statistical methods; (4)the definition fields and boundary condition of the units must be clear.

Generally, larger units are defined first and smaller units later. Comprehensive understanding of the signs of mineralization, correct securement of the premise of the prospecting project, and precisely selecting the statistical variables are the keys to correct division of units. As the units are of various scales, the sign and variables are of various scales, too. If the signs of mineralization in a unit are ambiguous, this unit is of no prospects. In fact, the division of units mirrors the inferences of the predictor about geological anomalies.

The units used in one metallogenic model are from the same kind of geological grosses, and have complete and unmixed compound of mineralizational sings. In each unit, the amount of mineral resource is fixed and can be figured out.

OPTION OF VARIABLES AND MATHEMATICAL METHODS

Geological variables are the statistical signs of marallotects extracted from the same kind of units. Selection of variables is based on the study of metallogenic laws and unit division. For the geological bodies are of different scales, the variables are of scales, too. The variables used in determination of the occurrence, estimation of the total amount, modeling metallogenic series, predicting large and/or superlarge deposits are different. The variables for different aim or in different stage of prognosis may not be the same, but one variable may exist in different stages.

Assignment of variables indicates the quantitative expression of geological variables of statistical units. The principles of assignment of variables are different for different prognosis aim and in different stage. Variables can be either quantitative or qualitative. The qualifications of variables decide what kind of mathematical techniques should be used. Some techniques are good at quantitative data processing, and some at qualitative data reckoning.

Ranking variables represent the scale of mineral resources and the concentrating degree of ore – forming element. They are used in quantitative estimation of mineral resources.

SELECTION OF MODEL UNITS

Model units indicate the units in which the reserves of mineral resources have good correspondence to metallotects. They are from the same gross, and have complete combination of prospecting indicators. The reserves in model units are of certainty.

The mathematical techniques include: (1)division of units, selection and assignment of variables;   (2)ranking the units according to their reserves;   (3)calculating the affiliation of the units and seeking for the overall distribution laws of the model units;   (4)separating the outliers, called non – standard units.

The causations of non – standard units include the existence of potential reserves, coming from different grosses and assignment errors. Non – standard units are noise for establishment of the prospecting models and must be abandoned. Only the units agreeable to the overall distribution laws are selected to establish the comprehensive prospecting model. The quantitative theory IV and V are used here.

ESTABLISHMENT OF PREDICTING MODELS

After the preparation of synthesization of information, research of metallogenic condition, selection of units, and option and assignment of variables, it is possible to appraise mineral resources quantitatively. The aim of prognosis decides the type of prospecting model and the corresponding mathematical tools. The algorithms used in QAMRII are not difficult, but the effect of prognosis depends on the user's understanding of the principle of QAMRII. The following explanations are essential to improve the result of prognosis.

(1). The model of statistical prognosis is established on model of units and unit assembles, and is used to locate the occurrences of prospecting area according to the theory of similarity – analogy. Multivariate statistical methods are used in quantitative data processing. Quantitative theories and characteristic analysis are used in qualitative and /or mixed data processing. Locating the prospecting areas is the basis of metallogenic prognosis.

(2). The model of mineral resources prognosis is based on the ordered variables assembles, and is used to estimate the amount of resources in each unit to be predicted. The corresponding techniques include logic information analysis and Monte – Carlo simulation.

(3). Prognosis of metallogenic series is to predict series of deposits related in metallogenetic time and space. It is applied to estimation the possibility of finding deposits of new types. It is emphasized to study the influence of geological variables to different kinds of deposits, which affect the characteristics of metallogenic series. The combination of variables controls the distributions of various kinds of 'deposits. The two – stage regression is used here.

(4). The theory of anomaly – seeking is used to forecast huge deposits. The order of units and variables are stressed. The establishment of the assembles of the extreme values is the key.

(5). To deal with the unbalance of information between model units and units to be predicted, metallogenic models and prospecting models are always simplified. To simplify the predicating models is to establish the model of information transformation.

THE EXPRESSION MODEL OF THE RESULT OF QAMRII

Application of artificial intelligence is necessary in mineral resource prognosis. This is demanded by either the complex processing of QAMRII or proper expression of the result. The process of QAMRII is realized by means of expert system now. The expert system can automatically output the necessary geological, geophysical, geochemical, and remote sensing charts and maps.

Neural networks have become the useful tools in QAMRII. Network for supervised model - recognition, such as BP network, is used to predict the deposits of known type. Network for non – supervised model recognition, such as Kohonen network, is used to predict the deposits of unknown type(1).

CONCLUSION

The theory of QAMRII is created and developed with the development of the theories of geology, economic geology, Geomathematics, and computer techniques. It is a result of the intersection of multiple sciences and technologies, and will impose the development of the intersection inversely.

REFERENCES

1. Chen Ming, Gold Resource Prognosis at 1: 500, 000 in Chinese Altay. [Unpublished Ph. D. thesis] . Changchun University of Earth Sciences, Changchun(1996), in Chinese with English abstract.
2. Chen Yuqi, Chen Yuchuan and Zhao Yiming, Preliminary discussion of minerogene – tic series of ore deposits, Journal of Academy of geological Sciences of China, 1,1(1979), in Chinese.
3. Chen Yuqi, Chen Yuchuan and Zhao Yiming, Second discussion of minerogenetic series of ore deposits, Journal of Academy of geological Sciences of China, 1,6(1983), in Chinese.
4. Wang Shicheng, Cheng Qiuming and Wang Yutian, Quantitative Appraisal Princi – ples and Methods of Mineral Resources, Journal of Changchun University of Earth Sciences, **16**, 15 – 24(1986), in Chinese with English abstract.
5. Wang Shicheng, Fan Jizhang and Yangyonghua, *Mineral Resources Appraisal*. Jilin Science – technique Press, Changchun (1990), in Chinese with English abstract.
6. Wang Shicheng, Hou Huiqun and Wang Yutian, *Study of Medium Scale Prognostic Methods for the Endogenic Deposit Series*, Geological Publishing House, Beijing. (1993), Chinese.
7. Zhao Pengda, Hu Wangliang and Li Zhijin, *Statistical Analysis of geological Exploration*, China University of Geosciences Press, Wuhan, (1990), in Chinese.
8. Zhao Pengda and Meng Xianguo, Preliminary discussion of geological anomaly, Journal of China University of Geosciences, **16**, 241 – 248(1991).
9. Zhao Pengda and Meng Xianguo, Geological anomaly and mineral prediction. Journal of China University of Geosciences, **18**, 39 – 47(1993)

Proc. 30th Int'l Geol. Congr., Vol.25, pp. 23-32
Zhao Peng-Da *et al* (Eds)
© VSP 1997

The Delineated Method of Geological Anomaly Units and Its Application in The Statistical Prediction of Gold Deposit of Large Scale

CHEN YONGQING, ZHAO PENGDA, and CHEN JIANGUO

Institute of Mathematical Geology & Remote Sensing Geology, China University of Geosciences, Wuhan ,430074, P.R.China

Abstract

The delineated method of geological anomaly units is the third one which comes after "the grid method "and "the geological body unit method ". Under "the geological anomaly ore-forming theory's guidance, it, based on the information extraction and synthesis of multi-discipline geodata, quantitatively delineates the units using comprehensive information. It is easy for the method to realize the automatic delineation of the units with the help of computer technique and keep an unit's information integrity at the same time.

Key words: geological anomaly ore-forming theory, delineation of geological anomaly units, comprehensive information, statistical prediction of gold deposit.

INTRODUCTION

The delineation of units is a fundamental work in the statistical prognosis of ore deposits. The prognostic results are based on the delineation of units. An unit, not only used as a statistical sample, but also as a carrier of mineral resources, is a joint which links the statistical model with the geological model. Therefore the delineation of units is an exceedingly important content in assessing ore capacity and directly related to the prognostic precision and effect.

At present, the methods of delineation units that are practicing in assessing mineral resources are outlined two categories: "grid unit method" and "geological body unit method"[1,2]. The key to the former is the determination of an unit size, however, which has no a definite rule. The later is a qualitative method, which doesn't facilitate to realize the automatic delineation of units. For this reason, we suggest the geological anomaly unit method.

"Geological anomaly unit method" means the delineated method of geological anomaly units. Under "the geological anomaly ore -forming theory's guidance, it, based on the information extraction and synthesis of multi-discipline geodata, quantitatively automatically delimits units using comprehensive information with the help of computer technique and keep an unit information's integrity at the same time.

GEOLOGICAL ANOMALY ORE-FORMING THEORY

Outline
In the middle of the 1970s, former the Soviet scholars Favorskaya and Tomson [3] regarded ore objects as geochemical anomalies associated with other kinds of geological anomaly. At the end of 1970s, former the Soviet scholar Gorelov [3] defined the geological anomaly from statistical view point that the geological anomaly means a statistically significant deviation in particular geological features of uniform geological objects in a given sector from the dominant background of the corresponding features of such object, during a given evolutionary stage of the structure as a whole. At the beginning of 1990s, Chinese scholar Zhao et al.[4] thought that the geological anomaly is a geological body or complex of geological bodies with obvious difference in composition and structure or orders of genesis as compared with its surroundings; the geological anomalous body that has a intrinsic relationship with ore-forming process is a necessary factor of mineralization and a indicator of ore deposits; which elucidated naturally the concept of the geological anomaly and the relationship between the geological anomaly and mineralization.

Ore-forming geological anomaly
Geological anomaly doesn't always bring about ore deposits, but is an essential prerequisite to the metallization[5]. The ore-forming geological anomaly means such a geological anomaly as has a intrinsic relation with metallogenesis and distribution of ore deposits. It is initially classified into two categories: ore -controlling geological anomaly and geological ore anomaly. The former means the geological factors controlled metallization and distribution of ore deposits, for example, ore-controlling strata, ore-controlling fractures, and ore-controlling intrusive bodies etc. The later means mineral resource bodies, for example, ore field, ore deposit, and ore body as well as their geochemical, geophysical, and remote sensing anomalies etc. The target of ore-forming prognosis on medium and small scale aims at ore field and mineralized zone, thus whole geological anomaly is the prognostic objective. The target of ore-forming prognosis aims at ore bodies and ore deposits on a large scale, thus only the geological ore anomalies is the prognostic objective. The controlled relationship between the geological anomalies on different scales and the corresponding levels of mineral resource bodies is an important research content of the geological anomaly ore-forming theory. It has been illustrated that the global geological anomalies control intercontinental metallogenic domains; the regional ones do metallogenic belts; the local ones do ore fields, ore deposits and ore bodies; micro-geological anomalies may be used to locate ore bodies[6].

Geological anomaly and geological complexity
Some scholars (Bogatskiy and Suganov, 1968; Borovko, 1971; Menaker,1976; Gorelov, 1979) [3]considered that the concept of the atypicality of geological structure of the ore fields must not be confused with the similar concepts of complexity, heterogeneity, and gradient of the geological environment. The concepts of "complexity" and "atypicality" are similar but not identical. The geological structures of the ore-bearing sectors are not simply (and not absolutely) complex, but are specific, anomalous, and distinct from their surroundings. The

most structurally complex major tectonic zones are rarely ore-enclosing. But here there is a problem of the scale effect. As far as the ore-forming geological anomaly units such the intercontinental metallogenic domains, and metallogenic province controlled by the global and the regional geological anomalies are concerned, geological anomalies are the same as the geological complexities .That is, the intensity of the geological anomaly is in direct proportion to the geological complexity. Large type of endogenic metallic ore usually coincides with the geological anomalies of high complexity. For example, the endogenic magnetite mineralization of the Kuznets Alatau in former the Soviet Union was shown to be spatially localized in regions of high complexity[3]. The known large types of the endogenic metallic ore deposits in China also are spatially localized in the geological anomalous regions with high complexity[6]. In size of ore body and ore deposits the complexity of geological anomaly doesn't spatially coincide with the geological ore anomaly (ore body and ore deposit). In Tengchong district, southwest China gold deposits (gold occurrences) are distributed at the margin of the geological anomalies measured by the entropy of fracture intensity[5] . The areas where geological structures are complex usually are the tectonic active areas. The multi-periodic tectono-magmatic activities may supply the passage ways and motive forces of moving up for ore liquids in the depth of the crust, but deposit of ore mass requires a relative stable space. The overwhelming majority of endogenic ore deposits don't usually exist within multi-periodic active deep faults and intersection points of multi-groups of faults, but exist in their secondary faults and nearby sectors of their intersections.

We have drawn inspiration from the mentioned above that construction of synthetic prognostic variables is mainly based on information of geological ore anomaly and ore-controlling geological anomaly. Ore-forming favorability not only relies on geological complexity but also on intensity of geological ore anomaly.

THE DELINEATED METHOD OF THE GEOLOGICAL ANOMALY UNITS

Basic characteristics
The characteristics of geological anomaly unit method are as follows:
1. Since geological anomaly is of the property of intensity, its units are able to be quantitatively delimited by its threshold to realize automatic delineation of units. Since geological anomaly is of spatial concept, the anomalous unit delimited by their threshold is a three dimensional body. Its mathematical characteristics of various kinds of ore-forming information and its varieties can be described by mathematical method to facilitate cubical evaluation of the units and three dimensional ore-forming prognosis.

2. Having obvious differences with its surroundings in texture and composition the geological anomaly is usually accompanied by geochemical, geophysical, and remote sensing anomalies etc. The former is the origin of the later; and the later is the reflection of the former in chemical as well as physical properties[7] . It is the methodological foundation which quantitatively delimits geological anomaly units using comprehensive information.

3. According to the sampling theorem the geological anomaly unit method meets the statistical conditions: a) The samples come from the same population; b) Random of sampling and

representative of samples are guaranteed. Since the units are quantitatively delimits with the same comprehensive information value by the same method guarantee the units coming from the same mother body . Since delineation of the units depends on ore-form favorability values, areas where the values are more than the threshold are all regarded as units. The delineation of units is not influenced in order of priority, that is, units maintain mutually independence. Since geological anomalous events are regarded to be random events, the random of sampling is warranted.

4. The units delimited by this method objectively reflect the geological features and distributed tendency of ore resource bodies . Since units are the specific position where ore bodies might exist and possible metallized areas thus guarantee a geological anomalous body's integrity. The units maintain high degree unity with their geological anomalies. Since every unit is a metallized sector with different metallized intensity whole distributed features of units reflect the spatial distributed tendency of ore resource bodies.

Delineated method

The delineated methods of geological anomaly unit have been exploring , which depend on the complexies such as multi-levels of geological anomalies and ore resource bodies as well as relativity of background and anomaly etc. Different scale of geological anomalies have different levels of information and contents. On ore-forming prognosis of medium and small scales, although an unit such a favorable ore-forming sector is different from its surroundings in some texture and composition, the interior of the unit is thought to be relative homogeneous, that is, any part within the unit is considered to has equal ore-forming probability. But on ore-forming prognosis of large scale this sector is regarded as heterogeneous as whole region both in texture and composition (Table1).It has been shown that there is a self-similarity among geological anomalies. For example, information features of the local geological anomaly of scale 1:50000 within 400 km^2 approximately reflect information features of the regional geological anomaly of 1:200000 within 40000 km^2 in gold metallized concentrated area of western Shandong province. The self-similarity of geological anomalies enlightens us that the unit concentrates ore-forming information of whole region is the best prospecting target area, and the ore-forming geological characteristics of whole region might be recognized by a detail synthetic research into the sector. Change of prognostic objective ,from ore-forming perspective area to ore body (or ore deposit), is also an important feature of large scale ore-forming prognosis. Since prognostic precision and accuracy are raised, prognostic difficulty is greatly increased ,but prognostic achievement may directly produce economic benefits.

According to the features of large scale ore-forming prognosis, the delineation of geological anomaly units should cover the following fields: a) under guidance of the geological anomaly ore-forming theory, "diagnostic" ore-forming geological anomalous information is extracted from multi-discipline geodata with the help of sophisticated information processing techniques; b) the information is effectively synthesized by appropriate mathematical methods; c) geological anomaly map is compiled using comprehensive information; d) geological anomaly units are delimited on the synthetic geological anomaly map by its threshold.

Table 1.   Comparison between the regional geological anomalies and the local ones of Tongshi gold fields
in western Shandong province

types	regional geological anomaly (40000 km², 1:200000 scale)	local geological anomaly (400km², 1:50000 scale)
ore-controlling geological anomalies	**1.** ore-controlling sector is of bi-layers of structures: uplayer is Precambrian meta-morphic green stone belt; downlayer is Cambrian-Ordovician carbonate rocks. **2.** ore-controlling tectonics are NW, EW, and NE trend of faults. **3.** ore-controlling intrusive bodies are Yan-shanian intermediate-acid and/or interme-diate alkalic complex.	1.favorable rocks are Au bearing hornblende and green schist series; dolomite, dolomitic limestone, and silicic limestone . 2.ore-forming tectonics are EW, NW and NW trend of faults **3.** gold ore bodies usually are accompanied by intermediate-acid (alkalic) veins.
geological ore anomaly	1.Leiwu skarn Fe-Au deposits (ore-con-trolling rocks are gabbro-diorite series.) **2.**Tongjing porphyry Cu-Au deposits (ore-controlling rocks are granite porphyry.) **3.**Tongshi hidden bombing altered breccia Au-Ag-Te deposit, skarn Fe-Cu-Au deposit(ore-controlling rocks are diorite porphyrite and alkalic syenite-porphyry)	1.porphyry Cu-Mo-Au ore spots. 2. small skarn Fe-Cu-Au deposits. 3. large hidden bombing altered breccia Au-Ag-Te deposit.

APPLICATION

Geological outline
The studied area is located at the southwest edge of Pinyi-Feixian tectono-volcanic basin in the
swell of western Shandong province, eastern China .Ore-controlling intrusive bodies are
Yanshanian (190 Ma-70Ma) alkalic complex emplaced in Archean metamorphic green rocks
and Palaeozoic carbonate rocks (Fig.1a).The complex consists of hypabyssal suites of diorite-
porphyrite and syenite-porphyry and are rich in K_2O (4.06-10.12%) and Au (5.40-12.46 ppb).
Ore-controlling strata are mainly Precambrian metamorphic green rocks and Cambrian,
Ordovician carbonate rocks. Main lithologies of the former are hornblende and green schist
series, being rich in Au (5.90-139 ppb), Cr, Ni, Co, V etc. Ones of the later are dolomite,
dolomitic limestone, and silicic limestone. Ore-controlling tectonics are EW-, NW-, and NE-
trending faults . The known commercial gold deposits (GL gold deposit) are situated within
hidden explosive breccia zone controlled by the EW- trending faults   about 1 Km east from
the alkalic intrusive complex. Ore minerals are native gold, electrum, and calaverite. Ore-
forming elements are Au, Ag, and Te .

Extraction and synthesis of ore-forming information
Data of the studied area cover a geological map of gold field on scale 1:10000, high resolution
magnetic data of scale 1:10000, and soil geological data of Au, Ag, Cu, Pb, Zn on scale
1:10000. Extraction and synthesis of information include ones of the single discipline and the

multi-discipline. This work must take the geological anomaly ore -forming theory as premise, the methodological principles of every subject as a rule, and the solving geological problem as purpose.

Information extraction and synthesis of the single discipline. High resolution magnetic data are usually used for inferring the concealed boundary of geological body, its three dimensional distribution, as well as the tectonic skeleton. Its essence is to reveal the intrinsic relationships among magnetic field, magnetic body, and geological body through the synthetic comparison analysis for them. The concrete sequences are as follows: a) Magnetic tectonic anomalous map is compiled in the basis of processing and interpretation for magnetic data; b) Overlap this map with geological map of the same scale, and compile geological-geophysical tectonic map through extracting the inferred fractures, and the concealed boundaries of geological bodies (strata and intrusive bodies) according to the magnetic field's features of the known fractures, and the exposed geological bodies. Soil geochemical data of scale 1:10000 are used for recognizing geological ore anomaly. Firstly, the data are standardized; secondly, determine relative association of ore-forming element through Factor analysis; thirdly, synthesize ore-forming elements into one value using their standardized data by mathematical formula (1).

$$y_i = (\prod_{j=1}^{m} x_j)^{\frac{1}{m}}, x_j \geq 0, i = 1,2,...,n; j = 1,2,...,m \qquad (1)$$

where y_i is a synthetic value, x_j is a standardized value of the associated element, n is sample number, m is the associated element number. Synthetic standardized geochemical map of the associated elements is compiled with y_i. The critical value that isolines vary from the sparse to the dense is regarded as the threshold of y_i, and delimit the associated element anomaly on the map by the threshold. Associated element anomaly more accurately reflects distribution of ore bodies than the single element. Ore -forming element association in studied area is Au-Ag-Cu[8].

Information synthesis of multi-disciplines. Information synthesis of multi-disciplines includes both qualitative and quantitative synthesis. The qualitative synthesis means that the extracted single discipline information of geology, geophysics, and geochemistry is synthetically expressed on one map with different symbols, forming the synthetic geological anomaly map(Figs.1a,1b. Fig.1a and Fig.1b are synthetized on their spatial coordinates to form the synthetic geological anomaly map). It wholly reflects geological, geophysical, and geochemical characteristics of the studied area. Quantitative synthesis means that various kinds of ore-forming information extracted from the synthetic geological anomaly map is respectively endowed with values, and the values are synthesized on a certain mathematical rule. Isoline map of the ore-forming geological anomaly is delineated   by the synthetic values. Qualitative synthesis is the foundation of quantitative synthesis.

Delineation of ore-forming geological anomaly units

Ore-forming geological anomaly units are delimited on the ore-forming synthetic geological anomaly map. The map is compiled by the synthetic values of ore-controlling geological anomaly and the associated anomaly of Au, Ag, and Cu .

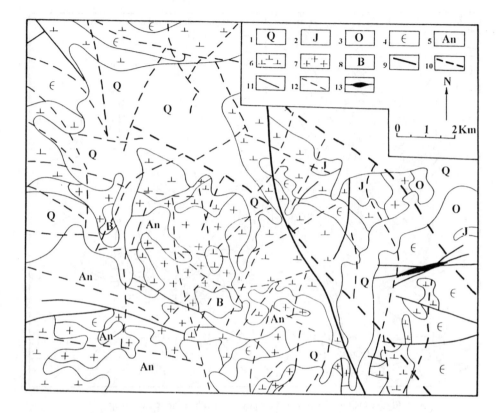

Figure 1a. Geological anomalous map of GL gold deposit. 1, Quaternary sediments; 2, Jurassic volcanic sedimentary rocks; 3, Ordovician carbonates; 4, Cambrian carbonates; 5, Archaean metamorphic rocks; 6, Yanshanian diorite-porphyrite; 7, Yanshanian syenite-porphyrite; 8, Hidden explosive breccia; 9, Deep seated faults; 10, Conjectured deep seated faults by high resolution magnetic data; 11, Shallow faults; 12, Conjectured shallow faults by high resolution magnetic data; 13, Gold ore bodies.

Information extraction and synthesis of ore-controlling geological anomaly. Since the geological anomalies of controlling location of ore bodies are some fractures and contact surface among geological bodies, size of fracture, number of crossed points of fractures in different directions and number of rocks within a unit area reflect the complexity of ore-controlling geological anomalies. The formula that calculates the complexity of ore-controlling geological anomalies is as follows:

$$C_x = \frac{1}{2}(n_1 + n_2)l_f \qquad (2)$$

where C_x is the complexity, n_1 and n_2 are respectively the number of crossed points of fractures and number of rocks within a unit area, l_f is total length of fractures within an unit area.

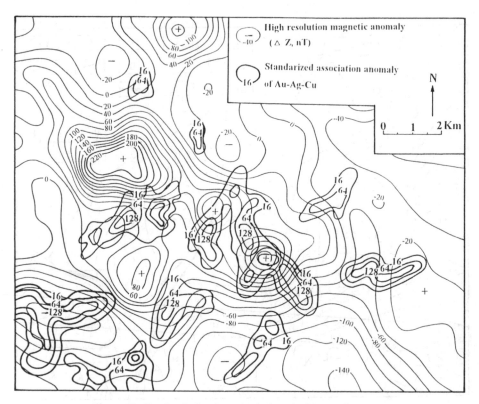

Figure 1b. Geochemical and geophysical synthetic anomalous map

Information extraction and synthesis of the associated element anomaly. That the highest value of the associated element anomaly is multiplied by its anomalous scope within a unit area is regarded as the anomalous intensity of the unit. Its calculated formula is as follows:

$$M_I = y_{max} \, S \qquad (3)$$

where M_I is the anomalous intensity, y_{max} is the highest value, S is the anomalous scope.

Compilation of ore-forming synthetic geological anomaly map. The formula of calculating ore-forming geological anomaly is as follows:

$$O_f = Ln \, (C_X + 1) + Ln \, (M_I + 1) \qquad (4)$$

where O_f is ore-forming favorability of the geological anomaly, C_x is the complexity of geological anomaly, M_I is the   anomalous intensity of the associated elements. The map of ore-forming synthetic geological anomaly is compiled by O_f values (Fig.2).

Take the value that the isolines vary from the sparse to the dense as the threshold of delimiting the units of geological anomaly on the map of ore-forming synthetic geological anomaly. For example, take 7 as the threshold, and delimit 5 units of ore-forming geological anomaly, among them the fifth unit is one where the GL gold deposit is situated (Fig.2).

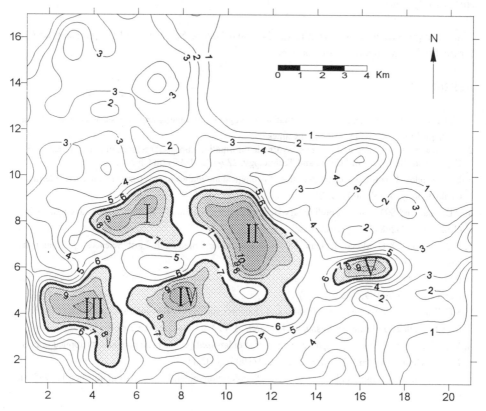

Figure 2. Ore-forming synthetic geological anomaly of GL gold field

CONCLUSION

"Geological anomaly unit method" is of a solid geological theoretical foundation, and is in conformity with statistical sampling theorem. In concrete sequence modern computing and image processing techniques are utilized and eventually automatically quantitatively delimit units using comprehensive information. Application of the method into large scale statistical prognosis of gold deposit achieves the desired results. Since there is a kind of corresponding controlling relationship between geological anomalies and ore resource bodies, the method also applies to assessing mineral resources of medium and small scale. However it has no limits for humanity to explore nature. Continual raising the accuracy and precision of ore-forming prognosis is a eternal subject of Mathematical Geology. Zhao [9]once pointed out that at present the main reason why perplexed precision and effect of ore-forming prognosis , in the final analysis, imputed to insufficient of information extraction and limitation of linear model. The effective way to solve the problem is to probe into   ways and techniques of extracting deep level of ore-forming information and establishing high resolution prognostic model. It is a subject to be further explored.

Acknowledgments

Sincere appreciation is expressed to the National Nature Science Fund Commission of China for the funding in support of this project.

REFERENCES

1. P.D. Zhao,   W.L. Hu and Z.J.   Li.   Statistical prognosis of mineral deposits, second edition. Beijing , Geological Press. 113-118 (1994) (in Chinese).
2. S.C. Wang, Q.M. Cheng and J.Z. Fan. Synthetic information appraisal methods of gold ore resources. Changchun, Jilin Press of Science and Technology. 229-232 (1990) (in Chinese).
3. D.A. Gorelov. Quantitative characteristics of geological anomalies in assessing ore capacity. Internal. Geology. Rew. **4**, 457-465 (1982).
4. P.D. Zhao and S.D. Chi.   A preliminary view on geological anomaly. Earth Sciences - J. China Univ. Geosci. **3** ,241-248 (1991) (in Chinese).
5. P.D. Zhao and X.G. Meng. Geological anomaly and mineral prediction. Earth Sciences - J. China Univ. Geosci. **1**,39-47(1993) (in Chinese).
6. P.D. Zhao, J.G. Wang and M.H. Rao et al. Geological anomaly of China. Earth Sciences - J.China Univ. Geosci. **1**, 97-106(1995) (in Chinese).
7. S.C. Wang and Y.Q. Chen.   Basic rules and characteristics of ore-forming series prognosis.Contributions to Geology and Mineral Resources Research. **4**, 79-85(1994) (in Chinese).
8. Y.Q. Chen and H.J. Jin.   Compiling method of standardized geochemical map and its applied effects. J. Changchun Univ. Earth Sciences. **2** ,216-221(1995) (in Chinese).
9. P. D. Zhao.   A certain number of new ideas of exploration mineral resources[unpublished].(1995) (in Chinese).

Proc. 30th Int'l Geol. Congr., Vol.25, pp. 33-42
Zhao Peng-Da *et al* (Eds)
© VSP 1997

Theory and Method of Phase-Separation Analysis of Remote Sensing Information Field of Metallogenetic Environment and Nonmodel Ore-Deposit Prediction*

YANG WUNIAN and **ZHU ZHANGSEN**
*Institute of Remote Sensing & GIS , Chengdu University of Technology ,
Chengdu 610059, P.R. China*

Abstract:

In this paper, the author proposes a new theory and method: the phase-separation analysis of remote sensing information field of metallogenetic environment and nonmodel ore-deposit prediction. By means of the theory and method, through decomposing and feature extracting of multisource information field of geological, geochemical and geophysical data based on RS, GIS and GPS techniques, a natural source model of ore-deposit can be established in order to predict the related deposits accurately. The theory and method have been practically applied in many tested areas and the effects are remarkable.

Keywords: remote sensing technique, mathematical geology, phase-separation analysis of remote sensing images, metallogenetic environment, nonmodel ore-deposit prediction.

INTRODUCTION

The traditional model prediction for ore deposits based on the analogic theory is mainly applied in the areas where sufficient geological study has been made and where deposit models can be easily constructed. However, it can only be used to look for the deposits those types and scales are similar to known deposits, but it is unsuitable to look for new type deposits or large/giant deposits. In this paper, the author proposes a new theory and method, the phase-separation analysis of remote sensing information field of metallogenetic environment and nonmodel ore deposit prediction. By means of this method and combining RS, GIS and other data, through decomposing of multisource information fields, feature extracting and seeking anomaly, a natural source model of ore deposit can be established in order to predict the related deposits accurately. The theory and method have been practically applied in many tested areas and effects are remarkable.

THEORY

The theory of nonmodel ore deposit prediction suggests that the forming of ore deposit result from

_____

* sponsored by The Open Research Laboratary of Quantitative Prediction, Exploration and Assessment of Mineral Resources, MGMR., China

multiple changes of many geological factors; and in a particularly geological environment, the places where tectonic movement of multi-times piled up together probably produce large or giant type deposits; and the existence of great mineralized body maybe lead to some remarkable differences in composition, structure, geophysical field and geochemical field from the surrounding geological background, even   lead to anomalies of   biosphere and atmosphere of the earth. Therefore, on the basis of geology and other data, combining RS with GIS, and through decomposing multivariate information fields, feature extracting and seeking anomaly, it is feasible to establish the natural models of ore source bodies to predict related ore deposits correctly. Since the method does not need any known deposit as the model for prediction, it can be used not only in new areas lacking of geological data,   but also in seeking new type deposits and giant deposits in old areas, so it is of momentous theoretical significance and practical value.

METHODS

Remote sensed image processing and information extraction

A remote sensing image is a stereoscopic miniature of natural landscape. It objectively records information on shapes and physical characteristics (tones and colors) of geology and structures both the individual parts and the overall patterns in mineralization environment,   and is a high quality summary. It also records some information on concealed structures. All of this information reflects the differences between geological bodies in composition, texture/structure, and physical properties under dynamic influences inside and outside. The information on concealed structures controlling ore can be revealed by deformation of the earth crust, and by modifications of the geophysical and geochemical fields, or by abnormal changes in the atmosphere and biosphere above them. Apart from such information, much information on remotely sensed multispectral images is invisible to the human eyes, such as infrared or microwave images. Through remote sensing image processing to extract indicating mineralization information not only provides useful data, but also is helpful for combining fracture systems of rocks and tectonic deformation with geological formations and metallogenetic processes very well, so as to achieve conclusions that coincide with objective reality.

Phase-separation analysis of remote sensing tectonic information fields

Geological and structural appearances displayed on remote sensing images are responsible to a composed result of the earth crust's movements since geological history, containing much indicating information of tectonic movements and related mineralization in different time and stages. Therefore, it is possible for us to decompose and extract related information from remote sensing images by means of the phase-separation analysis of remote sensing information field according to different time and structural layers, and then quantitatively describe and analyze the information with the geomathematic methods. However, most researchers have paid attention to the importance of time, scales, and space sizes of linear and circular structures in tectonic analysis and minerogenetic prediction, but there hasn't been an effective method to solve this problem. Many people often put these linear and circular structures in different time, scales, and structural layers together to quantitatively process, resulting in complexity of the problem, and it is very difficult to find out internal relations between linear and circular structures, related geological bodies, and related minerals. Final conclusion for minerogenetic prediction is uncertain. The authors proposed a new theory and method, Phase-separation analysis of remote sensing images for regional tectonic deformation and stress fields[4],   This method was used in many places in china and has achieved remarkable results[3 -10].

Mathematical decomposition of background and anomaly of remote sensing information fields

The main aim of mathematical analysis for linear and circular structural fields are as follows:
   A: to determine a boundary of background and anomaly;
   B: to extract minerogenetic information;
   C: to find out feature information of indicating minerogenetic processes.
Geomathematical methods, such as moving average, trend analysis, kriging, may be used.

Anomaly is relative to the background. As remote sensing structural information field, regional change trend of linear structures responses background, and these linear and circular features that reflect local structures are anomalies, such as concentrated belt of linear and circular structures. The background can be classified into single one and multiply one. For multiply background, it represents that the place passes more than one of geological processes, and variables of any geological factor may come from several maternal bodies. In this case, background can be determined using the phase-separation analysis method.

Multisource information processing of geological, geochemical and geophysical data with RS, GIS and GPS data

In the study of large-scale minerogenetic prediction, using the composite and alternate image processing of multisource information for a large quantity of geological, geophysical, and remote sensing materials, it can fast extract advantageous information and make some better representing for applications. However, the problem in this method is that generally used remote sensing images are rough corrected products which have some geometrical distortion. There are some displacements between geological bodies bearing-ore and their geophysical or geochemical anomalies in the 3-D space. If these were used to compose a map, a false compound anomaly map would be produced. A method to solve the problem is that an orthographic image map is made with RS, GIS and GPS techniques, and then multisource anomalies are precisely matched on the image map. 2- and 3-dimension multisource information maps are set up to quantitatively extract composite features to look for mineral resources. The author applied the method in the minerogenetic prediction for the Au in the tested area, Nanjiang, Sichuan, China, and achieved remarkable results.

EXAMPLES OF APPLICATIONS

Xichang Region

Xichang region, located in the northwest Sichuan, China, is complex in geology and tectonics. In the study of local structures bearing oil/gas and promising prediction, linear and circular structures were systematically interpreted with remote sensed TM images. Using the phase-separation analysis method of remote sensing information field, linear and circular structures in different scales and levels were quantitatively processed respectively[3, 6] .

(1). In the study of tectonic environment bearing oil / gas, through a quantitative processing of regional faults in large-scale, 2-dimension and 3-dimension quantitatively analyzing patterns of remote sensing information of regional tectonics were established (Pic. 1). Combining other data, several relatively static structural masses, where were fit to form petroleum and natural gas, were determined and as promising areas to look for oil/gas.

(2). For the selected areas above, with a special image processing, much indicating feature information was extracted (Pic.2). Related linear and circular structures were interpreted in detail. Using a combining quantitative analysis of circular structures and related radioactive lineaments (large-scale joints) combined with GIS and geophysical data, 2-D and 3-D quantitatively analyzing patterns of the local structures bearing oil/gas were established (Pics. 3, 4,) and much indicating information was extracted. Finally, several promising areas to look for oil/gas were determined.

Nanjiang region

Nanjiang region, located in the northern Sichuan, China, is complex in geology and rich in mineral resources. In this study, many geological and structural features were found through digital image processing and information extraction of Landsat TM images. Using geomathematical methods, these geochemical elements of Au, Cu, Pb, Zn, Fe, W and Co were processed. Applying a mapping technique of the orthographic image with RS, GIS and GPS, 1:50000 satellite TM orthographic image map was successfully made. With the map, 1:50000 interpreted geological map and linear / circular structural map were finished. Through composing geological, geophysical and geochemical anomalies with the orthographic image map, many indicating features were extracted and several Au mineralized points were found (Pics. 5 to 10).

REFERENCES

1. Zhu Zhangsen. About nonmodel prediction method. *Computer Techniques for Geophysical and Geochemical Exploration*, **14** (1),   60-62 (1992).
2. Zhao Pengda and Chi shuidu, On geological anomaly; *The Earth Science*, **16** (3), 241-248 (1991).
3. Yang Wunian, Zhu Zhangsen. The theory and method of parting-phase analysis of remote sensing information field and nonmodel ore-deposit prediction: *China Mathematical Geology (6)*, Geological Publishing House, Beijing, China. **6**, 67-71 (1995).
4. Yang Wunian, Yue Guangyu, et al., A new method for recovering regional tectonic stress field: *Science Bulletin*, Science Publishing House, Beijing, China,   36(12). 931-934 (1991).
5. Yang Wunian, Yue Guangyu, et al.,   Remote sensing image analysis for deformation fields and stress fields of the Jinfoshan rhomb-structure pattern in the contiguous zone of eastern of Sichuan and northern Guizhou: *Journal of Chengdu Institute of Technology*, **21** (1), 102-109 (1994).
6. Yang Wunian and li Yongyi et al., Applications of quantitative processing of remote sensing information to analysis of structures bearing oil/gas and prediction of promising areas in Xichang region, Sichuan, China. *Remote Sensing for Land & Resources*. **3**. 63-70. 1994.
7. Yang Wunian. Theory and method of parting-phase analysis of remote sensing images for regional tectonic stress field and deformation. In: Geological Society of China ed. *PROGRESS IN GEOLOGY OF CHINA (1993-1996) — Papers to 30th IGC*. Beijing: China Ocean Press. 1059-1063 (1996).
8. Yang Wunian. Remote sensing image analysis of tectonic patterns and structural stress fields of the liuzhi-Langdai region, western Guizhou: .*Remote Sensing for Land & Resources*. **28** (3).   21-28. 1996.
9. Yang Wunian, Zhu Zhangsen, Phase-separation analysis of remote sensing information fields and quantitative study of structural stress fields. *ACTA GEOLOGICA SINICA* (Chinese) **71** (1) (1997).
10. Yang Wunian. Phase-separation analysis of remote sensing information field, a new method for determining regional tectonic stress fields. *Proceedings of the 30th International Geological Congress*. VSP, International Science Publishers, The Netherlands. (1997).

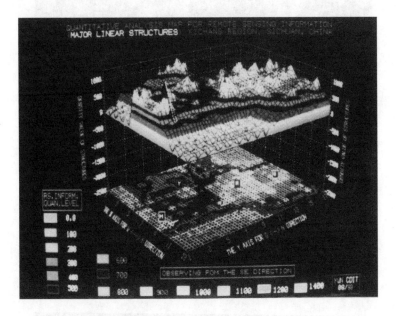

Picture. 1. 3-dimension color pattern of quantitative analysis for regional fault system in the Xichang region, southwestern Sichuan, China. The keys in colors from gray to white represent density values of regional major faults from 0 to 1440. The dark red areas in the below map are selected areas bearing oil/gas.

Picture. 2. Landsat TM4, 6 and 7 color composed image of Qiliba region, Xichang, southwestern Sichuan, China. The circular image feature in the middle of this picture reflects a local structure bearing oil/gas underground..

Yang Wunian   and   Zhu Zhangsen

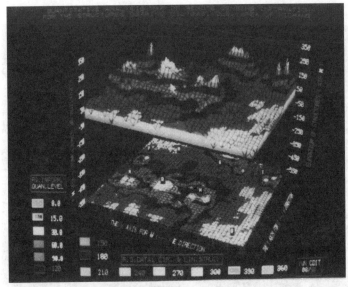

Picture. 3.

Picture. 4.

Pictures 3, 4.   3-dimension color patterns of quantitative analysis for local structures bearing oil/gas, Qiliba region of Xichang, southwestern Sichuan, China, observing from SE and NW directions separately. Quantitative data is information of circular structures and major joints related. Background values of remote sensing information of the structure development are 90-120. The anomaly values are 150 to 460. The higher anomalies (values >240) in EW-direction reflect information of an uplift zone of the concealed structures in EW-direction. The highest anomaly areas represent high points of concealed local structures bearing oil/gas. The solid maps upper part of the pictures represent appearances of local structures bearing oil/gas. The maps below the solid maps reflect distribution of local structures underground   5.5km. The color anomalies marked R (or RW), Q (or Q.B) or H are some concealed local structures bearing oil/gas.

Picture. 5. Landsat TM orthographic image map of Nanjiang region, Sichuan, China.

Picture. 6. Interpreted geological map with TM orthographic image map of Nanjiang region.

Picture. 7.   Interpreted linear and circular structural map with TM orthographic image map of Nanjiang region.

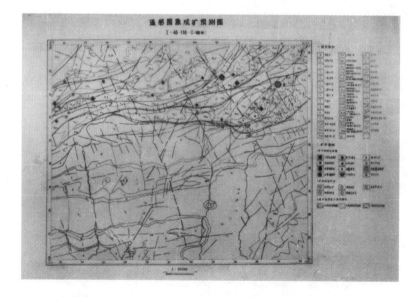

Picture. 8.   Metallogenetic prediction map with TM orthographic image map of Nanjiang region.

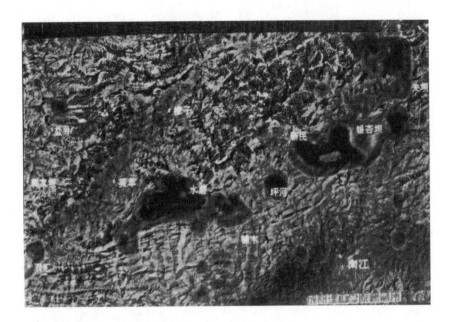

Picture 9. Composed map of gold   geochemical anomalies and Landsat TM 4, 5, 3 image, Nanjiang region.
Color features of gold anomalies from violet to red represent gold   geochemical abundance from lower to high.
Notices that these high geochemical anomalies of gold were controlled by a fault in NE direction.

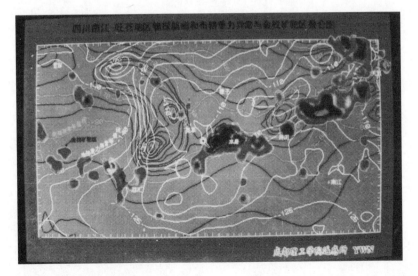

Picture. 10. Composed map of aeromagnetic anomalies, Bouguer gravity and gold geochemical anomalies (in red
color), Nanjiang region. It shows that Au mineralization were controlled by the big fault belt in NE direction.

Proc. 30th Int'l Geol. Congr., Vol.25, pp. 43-52
Zhao Peng-Da *et al* (Eds)
© VSP 1997

Phase-Separation Analysis Of Remote Sensing Information Fields, A New Theory and Method for Determining Regional Tectonic Stress Fields

YANG WUNIAN

Institute of Remote Sensing & GIS, Chengdu University of Technology
Chengdu   610059,   P.R.China

Abstract

This paper describes a new theory and method, Phase-separation Analysis of Remote Sensing Images for reconstructing regional structural deformation fields and recovering tectonic stress fields on the basis of characteristics of multispectral remote sensing images and   properties of the genetic relationship between major transverse tensional joints and related folds and faults. The main steps and procedures of the method are illustrated and some examples of applications are given.

Keywords: remote sensing technique, phase-separation analysis of remote sensing images, transverse tensional joints, linear structure, structural deformation field, tectonic stress field, method

INTRODUCTION

Tectonic stress field is important in the study of tectonics,   engineering geology, seismology and mineral exploration. Paleotectonic stress field traditionally has been determined mainly from information on rock   deformations and fractures collected from field outcrops. Because of the complexity of geological conditions, however, these traditional methods are limited by the discontinuity of observation points and lines in field, the availability of outcrops and access, and also by finances, so as often to lead to low efficiency and some suspicious results. Although much progress has been made in the use of remote sensing images to study Tectonics, the author found few papers concerning remote sensing   method for study of tectonic stress fields, and has not seen any mature report about such methods after reviewing the documents listed in GeoRef (1980-1995).

Some investigators think that, because remote sensing images   display mainly the last results of tectonic movements in the geological history of an area, the stage in a sequence of deformation and the time of tectonic deformation cannot be distinguished solely from remote sensing images and so it is impossible to solve the problems of tectonic stress field. What has happened here is that most researchers at home and abroad are still confined by traditional   theories   and methods of structural analysis. They interpret isolated lineaments or larger shearing faults in remote sensing images and determine a direction of principal stress according to the principle that the bisector of the conjugate angle of x-type   joints or faults is parallel to the direction of principal compressive stress. Without a systematic approach, and without using the entire joint system produced during tectonic deformation, it is impossible to assemble all of the relevant data on fractures related macro-structures   and relevant minerals as an entirety for analysis with remote sensing images.

it is also impossible under these conditions to make   quantitative statistics and analysis of tectonic stress   and deformation fields according to different scales, levels, and times from a picture of complicated tectonic deformation, not to mention   correctly assessing mineralization in associated with the different tectonic stress fields. Therefore, it is important both scientifically significance and for application to mineral prospecting to seek a new theory and method for determining regional tectonic deformation   and   stress fields.

A NEW METHOD FOR RECOVERING REGIONAL TECTONIC STRESS FIELD, PHASE-SEPARATION ANALYSIS METHOD OF REMOTE SENSING IMAGES

Theory, Method and Its Characteristics
The phase-separation analysis method of remote sensing images for regional tectonic stress field and deformation field is a means by which, macro-folds, related major transverse tensional   joints, major conjugate shear joints, and other fractures are systematically and phase ― separately interpreted with the remote sensing images   based on the dialectical relationship between remote sensing image factors and the phase-separation principle of geological interpretation of remote sensing images (Yang Wunian, 1986). The tectonic deformation fields are then determined through a comprehensive analysis of the macrostructures, the related fracture system , the distribution   of those joints in space, and their relation   to the tectonic deformation. Finally, according to the characteristics of   the tectonic deformation field, the structural stress states during different periods of rock deformation are reconstructed, and related stress fields are recovered. The method is mainly suitable for areas of buckle folding.

• Summary of geological structures, systematicness and continuity of   related information contained in remote sensing images

A remote sensing image is a stereoscopic miniature of natural landscape. It objectively records information on shapes and physical characteristics (tones and colors) of tectonic deformations, both the individual parts and the overall patterns, and is a high quality summary. It also records some information on concealed structures. All of this information reflects the differences between geological bodies in composition, texture/structure, and   physical   properties   under   dynamic influences inside and out. The information on   concealed structures can be revealed by deformation of   the earth crust, and by modifications of the geophysical and geochemical fields, or by abnormal changes in the atmosphere and biosphere above them. Apart from such information, much information on remotely sensed multispectral images is invisible to the human eyes, such as infrared or microwave images, and much of this information is useful in analysis of geological features out of sight. Using remote sensing images to analyze tectonic deformation and reconstruct tectonic stress fields not only provides useful data, but also is helpful for   combining fracture systems of rocks and tectonic deformation with geological formations very well, so as to achieve conclusions that coincide with objective reality.

• • A remotely sensed major joint represents a naturally preferred selection of the same group of structural joints in outcrop

Theories and modeling experiments have shown that in a fracture system of rock beds which was produced developed under pressure, each of the fracture groups always has a dominant orientation. The tectonic joints obvious on a remote sensing image are different from those observed at field outcrop. A remotely sensed major joint is a high-grade fracture belt in scale and orientation, which was initially produced along the dominant orientation of   the same group of joints observed in the field and which then was further developed through rock   deformation. Therefore, the major joints on a remote sensing image can well represent the same group of structural joints in the field, and are a natural preferred selection of the joints having dominant orientation. The features of trajectories and displacements of conjugate shear faults and major joints, as well as their relationship with main structures and principal axes of stress, are obvious on remote sensing images.

However, the major transverse tensional joints (perpendicular to fold axes and bedding surface) have not been used to full advantage seriously by investigators. Joints of this kind, in space, are distributed mainly on the two limbs of a fold and have a close relation to the fold geometry. They are distributed in parallel sets controlled by rock properties, and their orientations change with curves of the fold axes. In a remote sensing image, the joints are characterized by short lineaments paralleling one another, and show features sequence of gully-like landforms. In genesis, these joints are initially produced in a direction parallel to the maximum principal stress and are further developed during folding of the rock bed. They are commonly developed along the fracture trace of initial X-type conjugate shear joints, having stable properties and state. In mechanics, transverse tensional joints is parallel to the direction of the maximum principal compressive stress (σ 1) and perpendicular to the minimum principal compressive stress (σ 3). The direction of the intermediate principal stress (σ 2) is parallel to the surface of transverse tensional joints or to the line of intersection between the two crack surfaces of the conjugate shear joints. Thus, in a remotely sensed image the traces of transverse tensional joints can represent the axes line direction of the maximum principal compressive stress (σ 1). Using these relations, a network of principal stress in a study area can be easily mapped by interpreting and processing of remote sensing images. In light of mechanical properties of rocks, the density of joints developed in the same or similar rocks can, to some extent, reflect the relative magnitude of stress. So, through quantitative processing of data on joints, a quantitative map of stress distribution can be drawn.

••• Tectonic stress fields in a complicated area can be analyzed by phase-separation with remote sensing images based on the features of tectonic deformation

The structural pattern on a remote sensing image is a synthetic result of combination of the deformations produced during the geological history of the area, including many features of previous tectonic deformations. The tectonic stress fields of different stages in the complex deformation of an area can be reconstructed by the following steps:
(1) Geological structures from different stages and time are distinguished using remote sensing images;
(2) Tectonic deformation fields at different stages are determined by analyzing relationships between microstructures (joints and fractures) and the related tectonics (folds and faults); and
(3) Tectonic stress fields at different stages are respectively recovered through study of the features of structural deformation fields in different periods.

Theory and practice have shown that a specific tectonic form is produced in a specific structural stress field, and that in the specific stress field the structural deformation and associated joint system are closely related in space. If a folded area is subjected to again by a stronger force different in direction from the earlier one, some new folded folds with a new joint system will be produced. In such cases, the spatial disharmony of the folds and joints formed in different periods, as well as the changed characteristics of the earlier folds modified in the later stress field, can be detected on the remote-sensing image. Furthermore, since the early major transverse tensional joints were further developed along the initial joints perpendicular to rock-bedding surfaces, their properties might be changed under later stresses, and the rock blocks might be displaced along the joints. As a result, the orientation of the joints trajectory might be changed so that they are no longer perpendicular to the axes of the early fold. Such displacement, however, does not affect using of these joints because we use only the orientation of the joints trajectories in analyzing principal-stress axes. It is these feature and displacements mentioned above that provide us the evidence to distinguish multiple periods of tectonic deformation.

The Main Steps and Procedures of the Phase-Separation Analysis Method   of Remote Sensing Image for Tectonic Stress Fields

The overall steps and procedures of this phase-separation analysis method of remote sensing image for analysis of regional tectonic deformation   and stress fields is   shown in Figure 1.

EXAMPLES OF APPLICATIONS

Area of Complex Foldings

In the Liuzhi-Dayao area of western Guizhou Province, China, there are several   groups of folds, faults, and joint systems of different trends that form a complicated tectonic pattern. The mechanism of tectogenesis in this area has been debated for a long time. In this study, using the phase-separation analysis method of remote sensing information fields[2, 7],   geological and structural features   were systematically interpreted with remote sensing images (Fig. 2). Through an analysis of the combined folds, faults and related joint systems, three periods of structural deformation and the resulting deformation fields were determined. As a result, three   tectonic stress fields produced successively during the Yanshanian period were recovered (Figs. 3,4,5).

Tectonic Stress Field and Related Ore Deposits in the Langdai Area of Western Guizhou Province

In this study area, nine fold belts of different trends are mosaiced together to form a complicated tectonic pattern with the shape of a triangle. The three sides of the triangle consist of three anticlinal belts inside winding arcs. Along the bisector of the each angle of the triangular structure, a syncline is separately developed. Three more secondary anticlines or domes are developed on the borders, perpendicular to the axes of the three anticlinal belts. The major transverse tensional joints related to   the folds are widely developed, which are   parallel to each other and perpendicular to the axes of the folds, and have a trend of increasing   density of development from the center of the triangular structural pattern to its sides and vertexes (Figs. 6, 7).

According to features of the structural deformation field and the distribution of joints, a unified structural stress field was recovered (top right of Fig. 7). This stress field was formed during the Yanshanian under triangular boundary conditions formed by three big major faults and   under simultaneous lateral compressive forces from three directions. The magnitude of the stresses and the rock   deformation decreased progressively along arcuate trajectories from the sides and vertexes of the triangular block to its center (top right of Fig.   7). The folding and development of related fractures followed this stress network interpreted by the author, forming different folds and their associated joints in different parts of the stress field. The nine folds of different trends and scales were developed in relationship to one another, forming a coordinated and interrelated structural pattern that controlled the distribution of a series of mineral deposits. Thus, a structural stress field that controlled formation of   mineral deposits was provided, and some promising areas for Fe, Pb and Zn ore deposits and for natural gas were suggested.

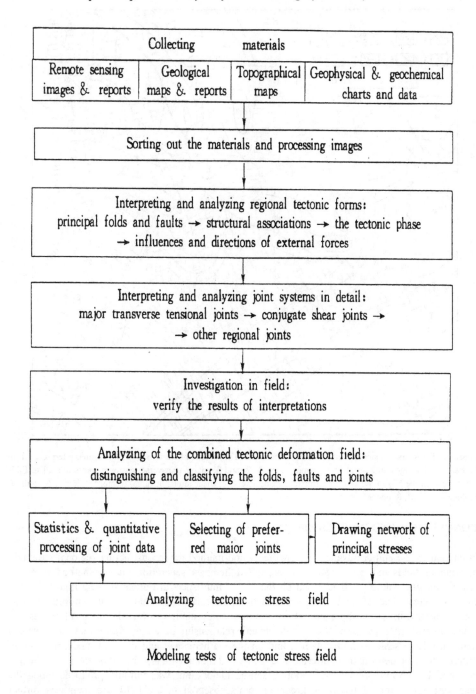

Figure 1. The main steps and procedures of this phase-separation analysis method of remote sensing image for analysis of regional tectonic deformation   and stress fields

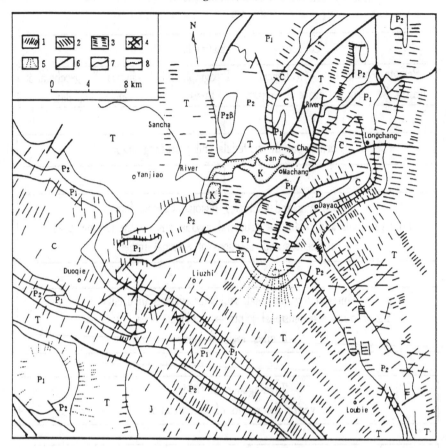

Figure 2. Transverse tensional joint systems, in the Liuzhi-Dayao area of western Guizhou, China, interpreted from Landsat MSS images and aerial photographs. Key: 1, 2, and 3, major transverse tensional joints related to folds having axes in the NW, EW, and NE directions respectively; 4, shear joints; 5, radial tensional   joints; 6, fault; 7, geological contact; 8, unconformability.

CONCLUSION AND DISCUSSION

In this paper, the author advances a new theory and method for reconstructing regional structural deformation fields and tectonic stress fields, called the phase-separation analysis method of remote sensing images. The method is based on the use of multispectral remote sensing images, which contain a great deal of geological and structural information (including much information on concealed structures), on the continuity of the image information to reveal geological bodies and structural properties, on the qualities of the genetic relationship between major transverse tensional joints and large-scale folds and faults, and on the mathematical and physical modeling tests of different tectonic deformation patterns. The theory and method are also based on the new principle proposed by the author that the remotely sensed major joint was initially produced along the dominant orientation of the same group of joints observed in the field and then was further developed through rock deformation, which provides a direction better measure of joint orientation than joints observed in field.   With this theory and method, the deformation and stress fields of several complicated tectonic patterns in Sichuan, Guizhou and Hubei provinces of China have been determined [2-9]. The studies provide relationships between tectonics, stress, rock deformation, and minerogenetic effects in the studied regions.

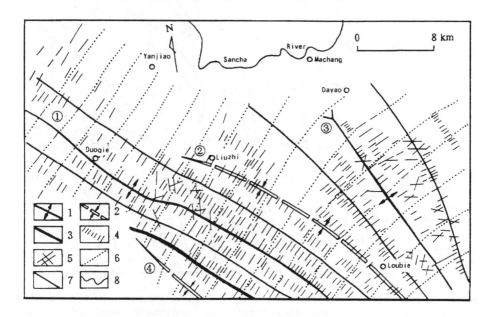

Figure 3. Stress distribution reflected by the first period of major transverse tensional joints. Key: 1, anticline axis; 2, synclinal axis; 3, fault; 4, transverse tensional joints; 5, shear joints; 6, maximum principal compressive stress (σ 1); 7, minimum principal compressive stress (σ 3).

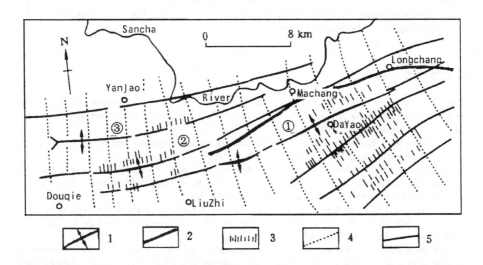

Figure 4. Stress distribution reflected by the second period of major transverse tensional joints. Key: 1, anticlinal axis; 2, fault; 3, transverse tensional joints; 4, maximum principal compressive stress (σ 1); 5, minimum principal compressive stress (σ 3).

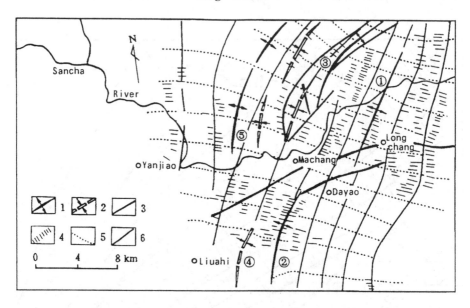

Figure 5.   Stress distribution reflected by the third period of major transverse tensional joints. Key: 1, anticlinal axis; 2, synclinal axis; 3, fault; 4, transverse tensional joints;   5, maximum principal compressive stress(σ 1); 6, minimum principal compressive stress(σ 3).

ACKNOWLEDGMENT

This paper   is a part of the achievement of the project sponsored by the national science foundation of China. The author wishes to thank sincerely Prof. Yue Guangyu and Prof. Zhu Zhangsen for their guidance and helps in the researching works.

REFERENCES

1.   Yang Wunian. Some dialectics in remote sensing geology and geological interpretation of remote sensing images: *Intersection-Sciences   and Natural Dialectics in Geoscience*. Management and Research of Geological Systems. 130-136 (1986).

2.   Yang Wunian and Yue Guangyu, et al. A new method for recovering regional tectonic stress field: Science Bulletin, Science Publishing House, Beijing, China. **36 (12)**, 931-934 (1986).

3.   Yang Wunian and Yue Guangyu, et al. Remote sensing image analysis for deformation fields and stress fields of the Jinfoshan rhomb-structure pattern in the contiguous zone of eastern of Sichuan and northern Guizhou: Journal of Chengdu Institute of Technology, **21 (4)**, 102-109 (1994)

4.   Yang Wunian and Zhu Zhangsen. The theory and method of phase-separation analysis of remote sensing information field and nonmodel ore-deposit prediction, *China Mathematical Geology (6)*, Geological Publishing House, Beijing, China. **6**, 67-71 (1995)

5.   Yang Wunian and li Yongyi et al., Applications of quantitative processing of remote sensing   information to analysis of structures bearing oil/gas and prediction of promising areas in Xichang region, Sichuan, China. Remote Sensing for Land & Resources. **3**. 63-70. (1994).

6.   Yang Wunian. Theory and method of parting-phase analysis of remote sensing images for regional tectonic stress field and deformation. In: Geological Society of China ed.: PROGRESS IN GEOLOGY OF CHINA (1993-1996)   Papers to 30th IGC. Beijing: China Ocean Press. 1059-1063   (1996).

7.   Yang Wunian. Remote sensing image analysis of tectonic patterns and structural stress fields of the liuzhi-

Langdai region, western Guizhou: Remote Sensing for Land & Resources. **28 (3)**, 21-28 (1996)..

8. Yang Wunian, Zhu Zhangsen, Phase-separation analysis of remote sensing information fields and quantitative study of structural stress fields. ACTA GEOLOGICA SINICA (Chinese) **71 (1)**, (1997).

9. Yang Wunian, Zhu Zhangsen. The theory and method of Phase-separation analysis of remote sensing information field of metallogenetic environment and nonmodel ore-deposit prediction. Proceedings of the 30th International Geological Congress. VSP, International Science Publishers, The Netherlands, (1997).

Figure 6.   Remote sensing image of the Langdai area (Landsat MSS 7, 5, 4, enhanced).

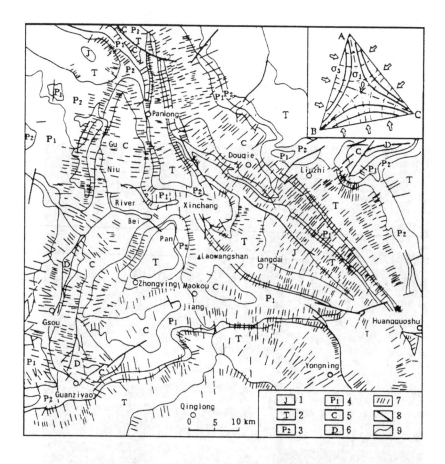

Figure 7.   Transverse tensional joint system and tectonic stress field (right top) of the Langdai area, interpreted from LANDSAT MSS images and aerial photographs. **Key**: 1, Jurassic; 2, Triassic; 3, Upper Permian; 4, Lower Permian; 5, Carboniferous; 6, Devonian; 7, major transverse tensional joints; 8, faults; 9, geological contact. Top right: dashed line, maximum principal compressive stress (σ_1); solid line, minimum principal compressive stress (σ_3);   arrows, lateral compressive forces.

Proc. 30th Int'l Geol. Congr., Vol.25, pp. 53-60
Zhao Peng-Da *et al* (Eds)
© VSP 1997

Mineral Reserve Estimation by ANN Technique in Termes of Neighborhood Information

LI Yuwei

Chinese Institute of Geology and Mineral Resources Information

LI Linsong

The First Computer Factory of Beijing

Abstract

Artificial Neural Network (ANN) technique is an attractive approach for pattern recognition. This paper describes a procedure to use ANN technique in reserve estimation in terms of neighborhood information instead of the information from the whole study area so that a successful result will be insured. A case study of application of this ANN approach to make block and point estimation of a highly variated and irregularly sampled gold deposit is given. The results are satisfactory and encouraging. A comparison between techniques of ANN and Kriging was made.

INTRODUCTION

There are essentially two approaches of estimating mineral reserves by spatial interpolation: inverse distance weighting method and Kriging. The former is easy to use but ignores geological variation. The latter takes account of not only the spatial configuration of samples but also the geological variation of a deposit. Some critical problems of Kriging applications, such as failure in building variograms, nonstationarity, etc., have been frequently encountered by geologists. Xiping Wu and Yingxin Zhou (1993) used Artificial Neural Network technique to estimate mineral reserves. This was a good application because ANN technique takes advantage of both inverse distance method and Kriging: easy to use and taking account of geological variation. In addition, when using ANN procedure for reserve estimation, one need not be concerned about problems such as variograms, stationarity, etc. Xiping Wu and Yingxin Zhou applied ANN approach to the whole study area in which only 48 sample points were studied. When the sample points increased, for instance more than 100, or the data configuration is complicated, it would be very difficult to reach a satisfactory ANN solution of reserve estimation. To improve the ANN approach for reserve estimation, the authors designed a new ANN procedure of reserve estimation which is based on the information within a neighborhood rather than that from the whole study area.

NEIGHBORHOOD DEFINITION

There are two reasons to use ANN technique for reserve estimation in a neighborhood instead of that in the whole study area. The first one is that the estimate of a block is only influenced by the samples in a neighborhood so that it is not necessary to determine the block value by using all of the samples in the whole study area. The second reason is that one may succeed in a ANN job of reserve estimation only with a small data set. When there are too many samples employed, an ANN iteration process may fail to converge. With a smaller data set, there will be a higher probability of success in convergence of ANN iterations. According to the authors' experience, an ANN job may run smoothly with samples less than 50. In short, we need a neighborhood to choose an effective data set from the complete one for a successful ANN operation to estimate a block value.

The definition of a neighborhood for ANN estimation is similar to that for Kriging. Initially, we design a grid system consisting of regular blocks. The problem here is to estimate the block value of a variable in terms of the observed values of samples in a neighborhood. For simplification, a neighborhood may be defined as a rectangle (Fig. 1). The samples within the neighborhood rectangle are employed for a block estimation. To determine the block value carefully, a few of estimate points are set up in a block being estimated. The ANN procedure returns a value at each estimate point of a block, the mean of which forms the block estimate (Fig. 2). This is quite similar to that for block Kriging.

The Neighbourhood Definition

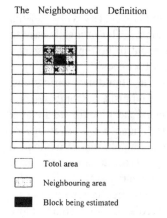

☐   Totol area

▦   Neighbouring area

■   Block being estimated

×   Sample points

The neighborhood size can be changed such that 50 sample points are included in it. There are essentially three factors influencing the accuracy of a block estimation. The first one is the data configuration in a neighborhood. An ANN estimation may be satisfactory if the sample points are evenly distributed. The second factor is the density of sample points. When a neighborhood contains more samples, the ANN procedure may return an estimate with a smaller error. The third factor is the spatial variation of the input variables. A highly fluctuated variable may cause large estimation errors.

Figure 1. Neighborhood definition.

THE ANN MODEL

Artificial Neural Network method is one of the most recently developed artificial intelligence techniques.   It has been successfully used in pattern recognition.   The attraction of this technique is that what a scientist applying ANN need to do is only analyse the input and output layers, or neurons. This means that people do not need to do anything about what happens in between, because   the hidden   layers are regarded as something   in a black box.   In this way,   people are   concerned about input and output affairs which are relatively easy,   while computers are used for the complex nonlinear relationship between inputs and outputs and do the tremendous computation jobs which are difficult. From this point of view, geologists using   ANN technique for reserve estimation may be spared many of the problems encountered when using Kriging.

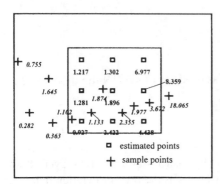

Block estimation is a process of spatial pattern recognition.   In consideration of the advantages of the ANN technique mentioned above and the previous paper by   Xiping Wu   and   Yingxin   Zhou(1993), we were encouraged to try this new pattern recognition approach for two dimensional reserve estimation.

Figure2. The samples and estimates

We designed a ANN structure including one input layer, two hidden layers and one output layer.   The input layer consists of three variables: easting (X), northing (Y) and the neighboring mean of grade.   So called neighboring mean here is the mean calculated by inverse distance method in terms of the four nearest samples.   Each of the two hidden layers consists of five neurons. There is only one output variable, the estimated ore grade.   This ANN structure is shown in Figure 3.

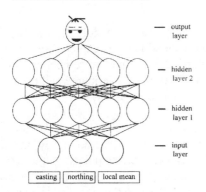

The ANN Structure for Grade Estimation

Figure 3. ANN structure

Let x_i be the input value of the i-th neuron of the lower layer, X_j be the primary output value of the j-th neuron of the higher layer, and w_{ij} be the weight between the i-th neuron in the lower layer and the j-th neuron in the higher layer. We have the following relationship between the j-th output neuron and the m input neurons

$$X_j = \sum_{i=1}^{m} w_{ij} x_i \qquad (1)$$

Then a characteristic function

$$x_j = \frac{1}{(1 + e^{-X})} \qquad (2)$$

is used to compute a proper output.

A simple back-propagation (BP) algorithm was used for this ANN iteration process. The weight adjustment equation

$$w(k+1) = w(k) + \eta \delta_j x_j \qquad (3)$$

takes the role of modifying the weights connecting neurons between two layers from the present value to the value of the next step; where k is the present iteration number, η is the learning rate, δ_j is the difference between the present value and the objective value of the j-th output neuron, and x_j is the output of the j-th neuron.

Although there are some modified BP algorithms and other more sophisticated learning algorithms, we used a simple BP algorithm as it worked well in our case.

A CASE STUDY

The ANN approach described above was used to study the Jingou gold deposit, Henna Province. This is a gold deposit of quartz vein type which was explored mainly by galleries associated with a few control drill holes. 250 samples were obtained from these exploration works. The samples are highly irregularly distributed and the gold grades are highly variated, as is apparent on the postmap of Figure 4.

A grid system with 25 by 9 blocks was designed. The size of each block is 50m by 50m. A block being estimated plus the surrounding blocks containing 50 sample points form an neighborhood. Each block contains 3×3 estimate points. In terms of these neighborhood parameters and sample data, we applied the ANN model defined above to estimate the gold grade block by block. The iteration number to reach a given accuracy varies in a wide range, depending on the number and configuration of samples in a neighborhood and the spatial variation of the variable employed. The critical iteration error was set to 0.001. When a real iteration error is less than this critical error, the iteration is finished successfully. Most of the iterations stopped at iteration numbers between 10000 - 30000. The maximum iteration number was up to 100,000. Each block was returned by the ANN procedure 3×3 point estimates of gold grades which formed the mean grade of the block. Both the point estimates and the mean of the block are important results of such an ANN grade estimation.

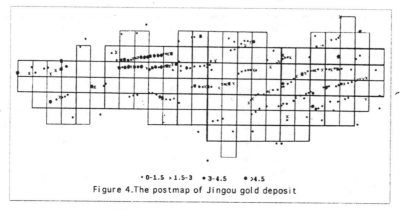

· 0-1.5 × 1.5-3 ● 3-4.5   ● >4.5
Figure 4.The postmap of Jingou gold deposit

Figure 4. The postmap of samples

Figure 2 shows one of the blocks of this gold deposit. The part of the neighborhood includes the nearest 11 sample points which are most important for the block being estimated. The 9 small squares represent the estimated points. It is noted that reasonable values were returned by the ANN procedure at these estimate points. An estimate point value was fitted to the nearest sample values very well. It is also noted from Figure 2 that the estimate at a point by this ANN approach seems to be determined only by a few of the nearest samples. The samples farther from that point might be negligible for the estimation. This is why we only need a neighborhood rather than the whole domain to evaluate a point or a block by the ANN approach.

The learning information of the neighborhood for the same block as Figure 2 shows is listed in Table 1.

After the weights were determined by learning, the easting, northing coordinates and the neighboring means of the estimate points of the block were into the expression (1) to compute the primary outputs. Then these primary outputs were substituted into the expression (2) to get the proper outputs. The point estimates of the block are listed in Table 2.

Figure 6 shows the block estimation and Figure 7 shows the point estimation for the whole deposit. It is noted that the point estimates in Figure 7 depict much more detail of the spatial feature of the gold grade. This is quite different from the estimates by Kriging. The Kriging results generally show little difference between maps of point and block estimates. The reason for this disparity is that Kriging usually makes very strong smoothing effects for both point and block estimation, but the estimates by ANN

Table 1. The Learning information of the block shown in Figure 2

Sample no.	Easting	Northing	Observed grade	Estimated grade	Error
19	18144.60	26058.27	1.645	1.114	0.531
20	18178.98	26056.50	1.847	1.868	-0.021
40	18206.20	26053.62	18.065	17.962	0.103
41	18199.19	26050.96	3.672	3.654	0.018
42	18190.78	26047.75	1.977	2.206	-0.229
43	18184.23	26045.26	2.355	2.206	0.149
44	18165.95	26046.87	1.133	1.155	-0.022
45	18158.85	26044.46	1.102	0.828	0.274
46	18147.42	26043.36	0.363	0.635	-0.272
47	18138.94	26042.65	0.282	0.605	-0.323
105	18130.54	26067.21	0.755	1.078	-0.323

Iteration times = 34828   S.D. = 0.2695

Table 2.   The point estimation of the block shown in Figure 2

Row	Column	Easting	Northing	Estimate
1	1	18162.73	26069.87	1.217
1	2	18179.40	26069.87	1.302
1	3	18196.07	26069.87	6.977
2	1	18162.73	26056.54	1.281
2	2	18179.40	26056.54	1.896
2	3	18196.07	26056.54	8.359
3	1	18162.73	26043.21	0.927
3	2	18179.40	26043.21	2.422
3	3	18196.07	26043.21	4.438

Block estimate = 3.202

do not cause such strong smoothing effects. Comparing the map of block estimates by ANN (Figure 6) with that by Kriging (Figure 5), although the global estimate by ANN (2.59375) is very close to that by Kriging (2.49658), it is apparent that the spatial feature of the estimate map by Kriging is much more smoothed.   Let us recall that geostatistics provides two models to study spatial data.   The first is Kriging model which carefully estimates the local means of a regionalized variable but does not take care of its spatial fluctuation.   The second is the conditional simulation model which carefully depicts the spatial fluctuation of a regionalized variable but does not take care of the local means very well.   The ANN procedure, for point estimation in particular, seems to incorporate both advantages of Kriging and conditional simulation.   An ANN point mapping procedure determines estimates at points whose spatial variation are very close to the spatial variation of samples.   It is obvious that the denser the sample points are, the closer to the real world the feature of the estimate map is.

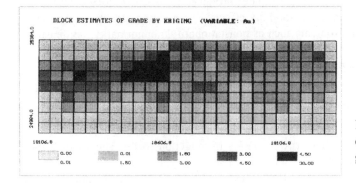

Figure 5.   Block estimates of gold grade by Kriging

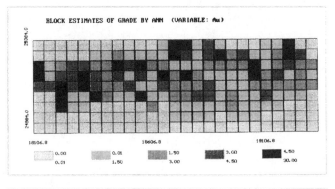

Figure 6. Block estimates of gold grade by ANN

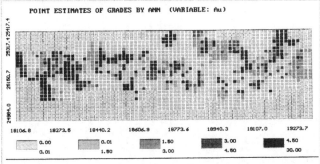

Figure 7. Point estimates of gold grade by ANN

To justify the ANN estimation, let us compare the ANN point estimates of Figure 7 with the postmap of samples of Figure 4.   Carefully examining the point estimates and their corresponding nearest sample values,   we find that they are accordant very well to each other.   This suggests that ANN may become a more attractive tool for mineral reserve estimation.

CONCLUSIONS

Artificial Neural Network technique is a new and effective approach for mineral reserve estimation.   An ANN job of local grade estimation defined in the whole study area may fail, but that defined in a neighborhood generally succeeds.   In comparison with geostatistics, the map of grade estimates produced by ANN is much closer to the real world due to the much less smoothing effects.   In addition, geologists no longer need be concerned about problems such as nonstationarity, nonnormality, nonlinearity, anisotropy, bad variograms, etc.   A critical problem for applications of ANN in mineral reserve estimation is that there is no way by ANN to calculate the estimation error.

REFERENCES

1. Hertz,J., Krogh,A., and Palmer,R.G., 1991, Introduction to the Theory of Neural Computation: Addison-Wesley Publ. co., Redwood City, California, 327p.
2. Xiping Wu and Yingxin Zhou, 1993, Reserve Estimation Using Neural Network Techniques, Computers & Geosciences, No.4, Vol.19.

Proc. 30th Int'l Geol. Congr., Vol.25, pp. 61-74
Zhao Peng-Da et al (Eds)
© VSP 1997

Optimization of Drilling Locations and Spatial Sampling Probability

ZHAO PENG-DA
Faculty of Earth Resources, China University of Geosciences (Wuhan), Wuhan 430074, P. R. China

WANG JIA-HUA
Department of Computer Science, Xi'an Petroleum Institute, Xi'an 710065, P. R. China

GAO HAI-YU
Faculty of Earth Resources, China University of Geosciences (Wuhan), Wuhan 430074, P. R. China
Present address: Department of Basic Sciences, Xi'an Petroleum Institute, Xi'an 710065, P. R. China

Abstract

Local probability of drilling an additional well, as the specialization of sampling probability when it is applied to selecting exploratory well locations in oil-gas exploration, is discussed. The model of integrating a variety of geological variables for calculating sampling probability, and an optimizing procedure for selecting drilling locations in oil-gas exploration, are proposed. The application results for a gasfield indicate the usefulness of this model and the optimizing procedure.

Keywords: exploratory well, dicovery well, optimizing procedure, oil-gas, the order of trap, the size of trap

INTRODUCTION

Usually, it is desired to select an exploratory well location in order to determine if a new area is oil-bearing in oil-gas exploration. This special well can be thought of as a discovery well once oil-bearing strata are met during the drilling procedures. In other word, discovery well is an exploratory well that locates a new oil-gas field or a new pool (reservoir). Moreover, once industrial oil-gas bed is found by discovery well in a pool, evaluating for the pool is usually followed in order to determine the probable extent of the reservoir. For this purpose, in general, delineation wells are required to design.

In a broad sense, there are some kinds of discovery wells, for example, a discovery well for an oil-gas basin (to be precise, a parametric well or an exploratory well that discovers oil-gas strata in the basin), a discovery well for an oil-gas field or a pool. Locating a discovery well, for an oilfield or a larger region, depends on the results of basin analysis and regional geological exploration. However, locating a discovery well, for a pool or a smaller area, can depend on the existing drillhole data, seismic data, and regional geological features only.

Estimating the most likely locations in an oil-gas field or a larger region where hydrocarbons are to be found has been widely discussed [6, 2, 3, 4, 8].

Locating the discovery well and delineation well, for a pool or a smaller area, will be discussed in this paper. In detail, the question is to choose some locations for additional wells in the area of interest, in which these locations might include the following features:

(1) *One location for discovery well*   It should be located in the most favorable oil-gas accumulation position. In general, the ridge or the high point of the structure could be the candidate for such a location.

(2) *Other locations for delineation wells*   A delineation well is such an exploratory well that is drilled after a discovery well to gain more information on the producing reservoir such as the elevation of the oil-water contact. A delineation well would be sited at a significant distance, usually two or more drilling and spacing units from the discovery well, but would be expected to become a producer.

Traditionally, the following procedures might be used: By eyes, compare a variety of maps interpreted from seismic data; then, artificially select exploratory well locations. The purpose of this paper is to model the traditional procedures as follows: By models, integrate a variety of geological variables; then, optimally select exploratory well locations.

In fact, the question at hand can be thought of as a sampling problem with unequal probabilities. So, it is most important to determine a model that can be used to calculate the sampling probabilities of spatial locations.

The another important part of sampling problem is the choice of a set of sampling locations that satisfies a given optimality criterion or procedure. This has been termed a search problem. Several search algorithms exist, for example, annealing algorithms [7], greedy algorithm and sequential exchange algorithm [1], two-phase algorithm [9], etc.

In this paper, local probability of drilling an additional well, as the specialization of sampling probability when it is applied to selecting exploratory well locations in oil-gas exploration, is discussed. A model for calculating sampling probability, and an optimizing procedure for selecting sampling locations in oil-gas exploration, are proposed.

LOCAL PROBABILITY OF DRILLING ADDITIONAL WELL

If an additional well is drilled within an area for some specific purpose, the probability that the drilling results obtained from the well can meet the given purpose can be termed the *local probability of drilling an additional well* (LPDAW) within the area. The larger LPDAW is within an area, the more favorable the area is for drilling an additional well.

It is clear that LPDAW is a function of the following factors.
    — The given area (position and size),

— The purpose of drilling an additional well,
— Varieties of geological characteristics (geological variables, sedimentary facies, etc.), and
— Economical, technical, and political factors.

As if the problem of selecting exploratory well locations can be thought of as a sampling problem with unequal probabilities, LPDAW can be thought of as the specialization of sampling probability when it is applied to selecting exploratory well locations in oil-gas exploration. In brief, LPDAW within an area can be also thought of as a measure for favorability of drilling an additional well within the area for the specific purpose. It is clear that, for the choice of discovery well locations of larger region, LPDAW is also a function of historical information on basinal evolution in relation to hydrocarbon production, migration, accumulation, and preservation.

Involved Features
It is notable that the problem is to optimally select the exploratory well locations, in which the area of interest is less than the degree of an oilfield. In such a situation, even if the basin assessment can be ignored, many factors are still involved in the above mentioned problem. At least, in the geological respects only, two different geological features can be envisioned.

(1) *Structure features* There are three most important structural factors:
— The order of trap,
— The size of trap, and
— Elevation and high point of structure.

(2) *Reservoir features* Two respects of physical property variables (parameters) of reservoir rock should be considered:
— Varieties of physical property variables of reservoir rock (seismic data), and
— The differences of the credibility of these variables.

The order of trap is illustrated here by using the following simple example. For a structural pool, for example, a simple anticlinal oil-gas trap, hydrocarbons are mainly dominated by the main anticline. Moreover, the anticline may dominate some sub-first grade structures. These sub-first grade structures may be the sub-first grade anticlines on the flank of the main anticline, or the fault blocks divided by main faults in the area of interest. So, in a broad sense, some traps exist in this example, such as main anticline, sub-first grade structures, or fault blocks. These traps, because of being located on the different positions of the main trap (anticline), have different favorability for drilling additional wells. This kind of favorability can be thought of as the *favorability of trap position*. The favorability of trap position is usually related to the depth of the trap top surface. The more shallow the trap top surface is, the more favorable the trap position is for drilling an additional well. It is in accord with the conventional geological exploration, because, in general, the ridge or high point of structure is the most favorable oil-bearing position.

The order of trap can be defined as the *magnitude* of the favorability of trap position for drilling an additional exploratory well. In this paper, the order of trap is generalized the magnitude of favorability of any subarea (not only trap) position for drilling an additional exploratory well. The calculation of the order of trap will be discussed in the next section.

Basic Model

As discussed above, for the problem of this paper, LPDAW is mainly related to the structural and reservoir features. For a subarea v, assume that S is the event that the structure characteristics within v are suitable for the oil-gas preservation, and R denotes the event that the physical properties of rock within v are not bad. So, the LPDAW within v can be defined as [5]

$$P(v) \equiv P_v(S\,R) = P_v(S)\,P_v(R|S) \tag{1}$$

In many petroleum regions, larger structural pool tends to be discussed early during exploration, because this kind of the pool is easy to be discovered. In such a situation, if the independence between structural characteristics and the physical properties of the reservoir is assumed to be met, one can get

$$P(v) = P_v(S\,R) = P_v(S)\,P_v(R) \tag{2}$$

The multiplicative relationship in equation (2) is naive and acceptable for exploration geologist, because hydrocarbons can be only preserved within a stratum where both the structure and the physical properties of the reservoir are not bad. The crux of the matter is how to provide a model for calculating $P_x(S)$ and $P_x(R)$ in equation (2), in which a variety of geological variables and the purpose of additional wells can be considered.

In cases where v is degenerated into a point x, and where only one of physical property variables of the reservoir is considered, Gao [5] proposed an approach for calculating $P_x(S)$ and $P_x(R)$, in which $P_x(S)$ is calculated by using the gradient of depth function of reservoir top surface, and $P_x(R)$ by using indicator function. In addition, the author neither discusses how to select the best locations for additional wells nor presents the application results for showing the suitability. Another model for calculating $P_x(S)$ and $P_x(R)$ in equation (2) is proposed in the next section. For simplicity only, LPDAW, $P(v)$, will be also called sampling probability in the following sections.

THE MODEL OF SAMPLING PROBABILITY

Based on equation (2), combining the structural and reservoir features in relation to selecting exploratory well locations, sampling probability, $P(v)$, within a subarea v, can be defined as

$$P(v) = K|v|s_o(v)s_s(v) \sum_{i=1}^{L} \mu_i \Pr\{\alpha_i \le z_i(x) \le \beta_i, x \in v\} \tag{3}$$

where

$z_i(x)$ is the value of ith physical property variable of the reservoir at point x,

$[\alpha_i, \beta_i]$ is the favorable interval of variable $z_i(x)$,

$\Pr\{\alpha_i \leq z_i(x) \leq \beta_i, x \in v\}$ is the probability that a random point x coming from v is subject to $\alpha_i \leq z_i(x) \leq \beta_i$,

weight μ_i indicates the credibility of variable $z_i(x)$, in which $\Sigma\mu_i = 1$,

$s_o(v)$ shows the order of trap for subarea v,

$s_s(v)$ shows the size of trap for subarea v,

$|v|$ is the measure of subarea v,

L is the number of physical property variables of the reservoir, and

K is a constant.

It is clear that K, $|v|$, $s_o(v)$ and $s_s(v)$ multiplied together, and the representation of summation sign in equation (3) correspond to $P_x(S)$ and $P_x(R)$ in equation (2), respectively. Assume that $s(x)$ is the depth function of the reservoir top surface. We consider each parameter of equation (3) in turn.

The *favorable interval of a variable* is defined as the interval in which the value of the variable on the favorable hydrocarbon area is almost contained while most values of the variable on the non-hydrocarbon area are not contained. Such a favorable hydrocarbon area can be roughly estimated according to the quantitative results obtained from basin analysis and seismic interpretation.

Weight μ_i, assigned to each physical property variable of the reservoir, indicates the credibility of the variable. Two meanings are implied in the credibility: one is the credibility of the variable because of the existence of errors in the measure or explanation for the variable; the other is the importance of the variable as a hydrocarbon indicator.

The *constant K* is considered so that $0 \leq P(v) \leq 1$ for every v. In fact, the constant can be ignored, because it does not have any influence on the relative value of $P(v)$. If $K=1$, the $P(v)$ in equation (3) can be termed the (sampling) *probability weight*.

The *order of trap* for subarea v, as discussed above, in addition to the value of $s(x)$, is related to the case whether a high point, maximum point in mathematics, on the surface of $s(x)$ is contained in v or not. Thus, one can set the following binary function:

$q(v)=1$ if v contains a maximum point of $s(x)$, $= r (<1)$ otherwise,

where r is a constant (in the application of this paper, we set $r=0.8$). Traditional methods for calculating maximum point by using partial derivatives can be used to determine whether a point is a maximum point or not. So, the order of trap can be defined as the multiplications between $q(v)$ and a linear decreasing function of the average depth of $s(x)$ within v, that is,

$$s_o(v) = q(v)\,[a - bd(v)] \tag{4}$$

where a and b are constants, and $d(v)$ is the average depth of $s(x)$ within v.

In order to determine a and b, one can set that $[a - bd(v)] = 0$ and 1 when $d(v)$ equals the deepest value and the most shallow value of the surface $s(x)$ on the whole area of interest, respectively.

The *size of trap* is an another structural feature. Because the reservoir top surface within oil-bearing trap should be convex upward, the size of trap is then defined as the probability that a random point coming from v is a convex upward point on the surface of $s(x)$, that is,

$$s_s(v) = \Pr\{x \in u_0 | x \in v\} \tag{5}$$

where $u_0 = \{x|\ x$ is a convex upward point on the surface of $s(x)$, $x \in v\ \}$. According to the rule of geometric probability, equation (5) can be written as

$$s_s(v) = \Pr\{x \in u_0 | x \in v\} = |u_0|/|v|$$

Define an indicator as

$$I(x,u) = 1 \text{ if } x \in u, \quad = 0 \text{ otherwise.}$$

Using this notation, one can get

$$s_s(v) = \Pr\{x \in u_0 | x \in v\} = |u_0|/|v| = \int_{x \in v} I(x,u_0)dx \Big/ |v| \tag{6}$$

The convexity or concavity of point for a function can be determined by using the first and second partial derivatives of the function respect to the components at spatial locations.

Similarly, set $u_i = \{x|\ \alpha_i \le z_i(x) \le \beta_i, x \in v\}$. Then the probability in equation (3) can be written as

$$\Pr\{\alpha_i \le z_i(x) \le \beta_i, x \in v\} = |u_i|/|v| = \int_{x \in v} I(x,u_i)dx \Big/ |v|, \quad i = 1,2,\cdots,L.$$

OPTIMIZING PROCEDURES

This section presents the *moving dividing* procedure of optimally selecting sampling locations in oil-gas exploration. This procedure can be stated as the following steps.

(1) The region of interest is divided into overlapped areas, v_i $(i = 1,2,\cdots,n)$, with the same size and shape (neglecting boundary effect) but different locations [v_i $(i = 1,2,\cdots,n)$ can be thought of as a set of moving neighborhoods]. Some of the areas, say $v_i^{(0)}$ $(i = 1,\cdots,m < n)$, are selected according to probabilities $P(v_i)$ calculated from equation (3) and the distances between the areas, in which m is greater than the predetermined number of exploratory wells to be selected.

(2) Similarly, each of $v_i^{(0)}$ $(i = 1, \cdots, m)$ is divided into overlapped subareas, and one of the subareas is selected like step 1. So, subareas $v_i^{(1)}$ $(i = 1, \cdots, m)$ are obtained.

(3) Repeat step 2 for $v_i^{(1)}$ $(i = 1, \cdots, m)$. Thus, small cells $v_i^{(k)}$ $(i = 1, \cdots, m)$ are obtained. (Maybe, some of the cells are equal.)

(4) Combining the geological interpretation, exploratory wells can be located.

In step 1 of the moving dividing procedure, $v_i^{(0)}$ $(i = 1, \cdots, m)$ are selected from v_i $(i = 1, 2, \cdots, n)$ by using the following approach.

Take $v_1^{(0)} \in V$ that is subject to

$$P(v_1^{(0)}) = \max_{v \in V} \{P(v)\},$$

in which $V = \{v_1, v_2, \cdots, v_n\}$. At stage $k+1$, take $v_{k+1}^{(0)} \in V$ that is subject to

$$P(v_{k+1}^{(0)}) = \max_{v \in V - V_k} \{P(v)\}, \quad \text{and} \quad d(v_{k+1}^{(0)}, v_i^{(0)}) \geq D, \quad i = 1, 2, \dots, k \qquad (7)$$

in which $V_k = \{v_1^{(0)}, v_2^{(0)}, \cdots, v_k^{(0)}\}$, $d(v_i, v_j)$ is the Euclidean distance between two center points of v_i and v_j, and D is a constant.

APPLICATION

Available Information
The area of interest is a gasfield in China. Available variables, reservoir top surface, porosity (both interpreted from seismic data), interval velocity, and frequency of seismic wave (figures 1 to 4) are obtained from a reservoir rock of the gasfield. Discrete values for these variables were measured at each center of 25 × 30 cells with a length of 200 meter and a width of 200 meter.

The question is to select three exploratory wells on the east of the western fault in figure 1 using equation (3) and the proposed moving dividing procedure. One of them is thought of as a discovery well, and the others as delineation wells for the gasfield. The values at cell centers on the west of the western fault cannot be used because of remarkable displacement of the fault. Based on the results obtained from basin analysis and seismic interpretation, table 1 shows the favorable intervals of the last three variables and related weights to be used in equation (3).

Table 1   Favorable Intervals and Weights to Be Used in Equation (3)

Variables	Favorable Intervals	Weights
Porosity	[7.0, 100.0]	0.5
Frequency	[35.0, 41.0]	0.3
Interval velocity	[3500.0, 3700.0]	0.2

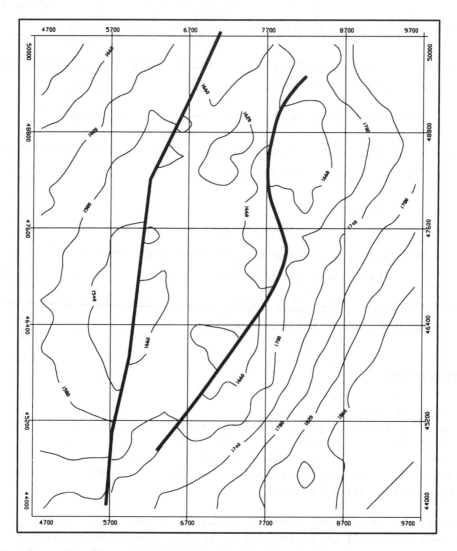

Fig. 1 Contour Map of Reservoir Top Surface. The bold polygons are the faults.

Dividing Approach

During the moving dividing procedure, we set $D=800$ in equation (7), and construct circular neighborhoods with 800 meter radius,

$$v_{i+1\,j+1} = \{x = (p, q)|\ (p\text{-}6100\text{-}400i)^2 + (q\text{-}45200\text{-}400j)^2 \leqslant 800^2\ \},$$

$$i = 0,1,...,7;\ j = 0,1,...,10.$$

Ten circular neighborhoods selected with 800 meter radius after the first step of the moving dividing procedure are shown in table 2.

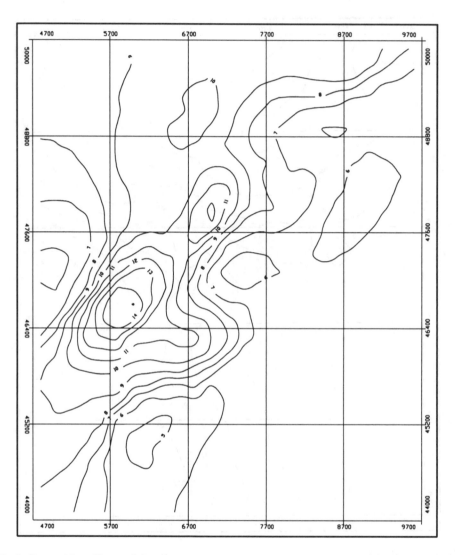

Fig. 2 Contour Map of Reservoir Porosity.

Let (s, t) denote the center point of $v_i^{(l)}$. In the second and third step of the moving dividing procedure, each $v_i^{(l)}$ (l=0,1,2) is divided into the following overlapped subareas:

$$\{x = (p, q) \mid (p\text{-}s\text{-}f\text{-}200ig)^2 + (q\text{-}t\text{-}f\text{-}200jg)^2 \leqslant w^2\}, \quad i, j\text{=}0,1,\dots,h$$

where

$$w\text{=}400 \times 2^{-l},$$
$$f\text{=}200 \times [0.006w]\text{=}200 \times [2.4 \times 2^{-l}],$$

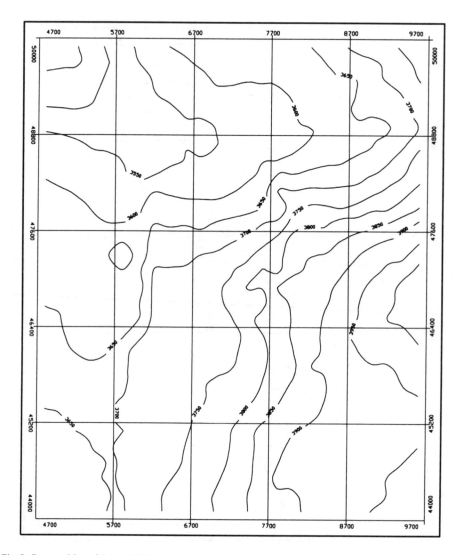

Fig. 3 Contour Map of Interval Velocity.

$g=2$ if $w \geqslant 400$, $=1$ otherwise,

$h=\max \{[0.01fg^{-1}]-1, 0\}$,

$[z]$ denote the integer part of digit z,

$l=0,1,2$.

Table 3 shows the results after the third step.

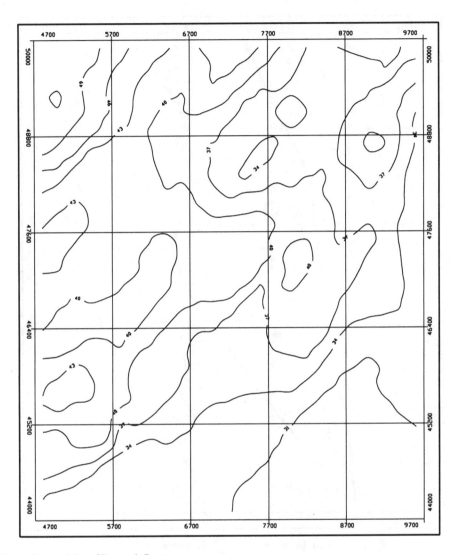

Fig. 4 Contour Map of Reservoir Frequency.

Numerical Results

According to figure 1 and tables 2 to 3, within 100 meter position error, we can infer that W1=(6700, 47800), W2=(7900, 48600), and W3=(6900, 46400) as shown in figure 5 are the optimized drilling locations. First, sampling probability weights within the subareas in table 2 and cells in table 3 associated with locations W1 and W2 are greater than most of that associated with the others. Then, the drilling of these two locations could have geologist get the information coming from different fault blocks (see figure 5). Last, the

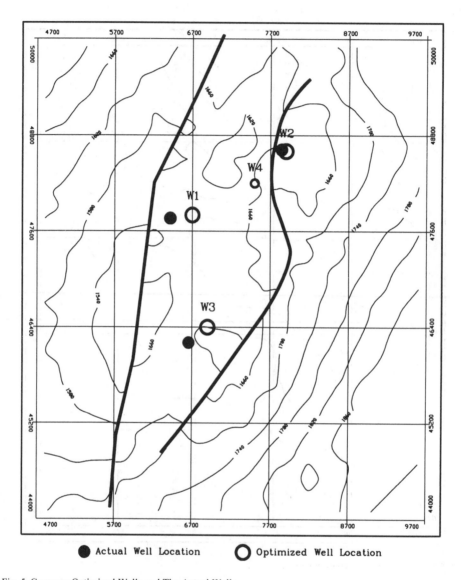

● Actual Well Location   ○ Optimized Well Location

Fig. 5 Compare Optimized Wells and The Actual Wells.

distance between each pair of the three locations is suitable. Figure 5 shows that the
optimized locations are near to the actual well selected by geologist. So, the approach
discussed here is useful. Moreover, W4=(7500, 48200) and W3 might be the suitable
drilling locations if without the eastern fault.

Table 2. Ten Circular Neighborhoods with 800 Meter Radius from the First Step of Moving
Dividing Procedure [1]

No.	Centers of Neighborhood	$P_1(v)$ [2]	$P(v)$
1	7300.0, 48000.0	0.551	27.011
2	8100.0, 48400.0	0.439	21.495
3	7300.0, 48800.0	0.398	19.512
4	6500.0, 48000.0	0.447	19.235
5	6500.0, 47200.0	0.372	17.869
6	7700.0, 46000.0	0.294	14.399
7	8100.0, 47600.0	0.293	14.368
8	7300.0, 47200.0	0.293	14.345
9	6900.0, 45600.0	0.293	14.341
10	6900.0, 46400.0	0.264	12.937

[1]   The values of $P(v)$ in the table are sampling probability weights.

[2] $P_1(v) = P(v)/|v|$.

Table 3. Eight Circular Neighborhoods with 100 Meter Radius Related to Table 2 after the Third Step of
Moving Dividing Procedure

No. (Related to the No. of table 1)	Centers of Neighborhood	$P(v)$	Order of $P(v)$
1,3	7500.0, 48200.0	0.537	3
2	7900.0, 48600.0	0.495	5
4,5	6700.0, 47800.0	0.541	2
6	7500.0, 45400.0	0.303	8
7	7900.0, 47800.0	0.496	4
8	6900.0, 47600.0	0.555	1
9	6700.0, 45400.0	0.363	7
10	6900.0, 46400.0	0.393	6

DISCUSSION AND LIMITATION

The model of sampling probabilities in equation (3) is slightly simplistic for many real problems. For example, sedimentary facies should be considered in the choice of exploratory well location in application. However, the approach presented in this paper provides a framework under which more complicated models can be developed.

The approach here is based on the hypothesis of structural pool. The model of sampling probabilities should be appropriately modified if the pool is controlled by lithology.

The model of equation (3) can be simply generalized to the situation when available information comes from three-dimension space by weighted sum of sampling probabilities within the subareas that are located in the reservoir rocks of different depths.

CONCLUSION

Locating the exploratory well locations in geological application, as a sampling problem with unequal probabilities, is quite complicated. The problem is how to integrate a variety of geological variables (parameters) in order to provide an appropriate model that can be used to calculate the local probabilities of drilling additional wells in oil-gas exploration. A heuristic model is proposed in this paper for selecting new exploratory wells. The application results for a gasfield indicate the usefulness of this approach.

REFERENCES

1. D. Aspie and R.J. Barnes. Infill-Sampling Design and the Cost of Classification Errors, *Math. geology* 22, 915-932 (1990).
2. E. Barouch and G.M. Kaufman. A Probabilistic Model of Oil and Gas Discovery. In: *Methods and Models for Assessing Energy Resources*. Michel Grenon (Ed.). pp. 248-260. Pergrmon Press (1979).
3. C.F. Chung and F.P. Agterberg. Poisson Regression Analysis and Its Application. In: *Quantitative Analysis of Mineral and Energy Resources*. C.F. Chung, et al. (Eds.). pp. 29-36. D. Reidel Publishing Company (1988).
4. J.C. Davis. Statistical Evaluation of Petroleum Deposits before Discovery. In: *Quantitative Analysis of Mineral and Energy Resources*. C.F. Chung, et al. (Eds.). pp. 161-186. D. Reidel Publishing Company (1988).
5. H.Y. Gao. Design of Discovery Well Location and Sampling Probability in Oilfield Exploration. In: *New Advance in the Theory and Application of Petroleum Science and Technology* (in Chinese). Proceedings of the Science and Technology Conference of Xi'an Petroleum Institute. R.Z. Lin, et al. (Eds.). pp. 112-116. Shaanxi Science and Technology Publishing House (1996).
6. G.M. Kaufman, Y. Balcer and D. Kruyt. A Probabilistic Model of Oil and Gas Discovery. In: *Methods of Estimating the Volume of Undiscovered Oil and gas Resources*. Studies in geology No.1, American Association of Petroleum geologists. J.D. Haun (Ed.). pp. 113-142. Tulsa, Oklahoma (1975).
7. J. Sacks and S. Schiller. Spatial Design. In: *Statistical Decision Theory and Related Topics IV* 2. S.S. Gupta and J.O. Berger (Eds.). pp. 385-399. Springer-Verlag, New York (1988).
8. A.R. Solow. On Assessing Dry Probabilities in Offshore Oil and Gas Exploration: An Application of Bayes's Theorem. In: *Quantitative Analysis of Mineral and Energy Resources*. C.F. Chung, et al. (Eds.). pp. 187-198. D. Reidel Publishing Company (1988).
9. D. Veneziano and P.K. Kitanidis. Sequential Sampling to Contour an Uncertain Function, *Math. Geology* 15, 387-404 (1982).

Proc. 30th Int'l Geol. Congr., Vol.25, pp. 75-88
Zhao Peng-Da et al (Eds)
© VSP 1997

Identification Probability and Pseudo-Entropy Criterion to Locate Drilling Locations

GAO HAI-YU
Faculty of Earth Resources, China University of Geosciences (Wuhan), Wuhan 430074, P. R. China
Present address: Department of Basic Sciences, Xi'an Petroleum Institute, Xi'an 710065, P. R. China

ZHAO PENG-DA
Faculty of Earth Resources, China University of Geosciences (Wuhan), Wuhan 430074, P. R. China

WANG JIA-HUA
Department of Computer Science, Xi'an Petroleum Institute, Xi'an 710065, P. R. China

Abstract

An approximate model, based on Bayesian law, for calculating identification probability of classification problem is presented. An encoding procedure for existing variables, in which the geologist's adventurous psychology and the credibility of the original data can be considered, is discussed in order to integrate spatial data. A pseudo-entropy criterion is proposed in order to select optimally drilling locations in oil-gas exploration. A case study based on the modified data of a gasfield is included.

Keywords: drilling location, appraisal well, oil-gas, classification, encoding function, Bayesian law

INTRODUCTION

Once industrial oil-gas bed is discovered in a trap in oil-gas exploration, evaluating for the trap is usually followed in order to estimate or determine the pool limits. For this purpose, in general, appraisal drilling, particularly, appraisal well locations are required to design.

In a broad sense, designing drilling locations can be thought of as a spatial sampling problem with unequal probability. A significant quantity of work has been published on spatial sampling design [6, 11, 10, 5, 7]. Spatial sampling in earth sciences is often related to the techniques of integrating and classifying spatial data. Specially, it is important to embrace suitable integrating and classifying techniques for the design of drilling locations, because of the complexity and the high costs in oil-gas exploration. For this reason, classification or integration problem has been widely considered [14, 9, 8, 2].

The most widely used criterion in the literature for sampling design is to minimize some kinds of the estimation variance for the area of interest [3, 7]. The criterion of minimizing the cost of classification errors [1], and the criterion of minimizing the chance of areas where values are significantly different from predicted values [12] are also discussed.

This paper discusses the following problem. Based on a variety of existing geophysical variables in oil-gas exploration, propose an appropriate model to select a set of new appraisal drilling locations in order to determine the pool limits. It is assumed that the extent of the area of interest corresponds to an oil-gas trap. Based on the Bayesian law, an approximate model for calculating identification probability of classification problem is presented. The encoding of existing variables is discussed in order to integrate spatial data and calculate identification probability. The encoding procedure can represent the geologist's adventurous psychology and the credibility of the original data. A pseudo-entropy criterion is proposed in order to select optimally drilling locations. A case study using a set of the modified data of a gasfield is included.

IDENTIFICATION PROBABILITY OF CLASSIFICATION

Assume that the area of interest can be divided into two categories, A and $\overline{A}$. $D(x)$ is an indicator function, that is,
$$D(x)=1 \text{ if } x \in A , \quad = 0 \text{ otherwise}$$

Usually, the classification of the area should depend on some geological characteristics. In general, suppose that *geological environment variables* $g_i(x)$ $(i = 1,2,\cdots,v)$ are related to the classification of the area, in which x is a spatial point. The reason of terming $g_i(x)$ geological environment variables is that these variables may not be obtained directly as normal geological variables but obtained by analyzing, calculating, or integrating procedure. In addition, geological environment variable represents integrative geological characteristics in some aspect. For example, $g_1(x)$ may represent integrative oil-forming conditions, $g_2(x)$ represent integrative structure conditions, $g_3(x)$ represent integrative reservoir conditions, etc. $g_i(x)$ $(i = 1,2,\cdots,v)$ may be discrete or continuous, and $G_i(x)$ are the random functions associated. Set
$$G(x) = [G_1(x),G_2(x),\cdots,G_v(x)]^T, \quad g(x) = [g_1(x),g_2(x),\cdots,g_v(x)]^T$$
Geological environment variables are not required to be obtained under some specific linear hypothesis as we can see in this section below.

Dividing the area of interest into A and $\overline{A}$ is a classification problem. The reasonable approach of dealing with this problem is to estimate the following conditional probability:
$$p_A(x) \equiv P[D(x) = 1|G(x) = g(x)] \tag{1}$$
Equation (1) is without meaning if $P[G(x)=g(x)]=0$, So, for given x, if the following limit is in existence:
$$P[D(x) = 1|G(x) = g(x)] = \lim_{\substack{\varepsilon_i \to 0^+ \\ i=1,2,\cdots,v}} P[D(x) = 1| \ |G_i(x) - g_i(x)| \le \varepsilon_i, i = 1,2,\cdots,v] \tag{2}$$
$P[D(x)=1 |G(x)=g(x)]$ can be termed the *generalized conditional probability* of $\{D(x)=1\}$ under the condition $\{G(x)=g(x)\}$, and $p_A(x)$ in equations (1), as viewed from

application, can be termed the *identification probability* of individual x being assigned to category A.

From Bayesian law, one can get

$$P[D(x) = 1 \mid |G_i(x) - g_i(x)| \le \varepsilon_i, i = 1,2,\cdots,v]$$

$$= \frac{P[D(x) = 1] \cdot P[|G_i(x) - g_i(x)| \le \varepsilon_i, i = 1,2,\cdots,v \mid D(x) = 1]}{P[|G_i(x) - g_i(x)| \le \varepsilon_i, i = 1,2,\cdots,v]} \tag{3}$$

Assume that $g_i(x)$ $(i = 1,2,\cdots,v)$ are continuous. If ε_i $(i = 1,2,\cdots,v)$ are very small, two equivalent infinitesimal relationships can be obtained:

$$P[|G_i(x) - g_i(x)| \le \varepsilon_i, i = 1,2,\cdots,v \mid D(x) = 1] \sim$$

$$2^v \varepsilon_1 \varepsilon_2 \cdots \varepsilon_v f^{(D)}[g_1(x), g_2(x), \cdots, g_v(x)]$$

$$P[|G_i(x) - g_i(x)| \le \varepsilon_i, i = 1,2,\cdots,v] \sim$$

$$2^v \varepsilon_1 \varepsilon_2 \cdots \varepsilon_v f[g_1(x), g_2(x), \cdots, g_v(x)] \tag{4}$$

where $f[y_1, y_2, \cdots, y_v]$ is the multivariate probability density of $G_i(x)$ $(i = 1,2,\cdots,v)$, and $f^{(D)}[y_1, y_2, \cdots, y_v]$ is the multivariate probability density of $G_i(x)$ $(i = 1,2,\cdots,v)$ under the condition $\{D(x)=1\}$.

From equations (1) to (4) and $p_0(x) \equiv P[D(x) = 1]$, one can get:

$$p_A(x) = \frac{P[D(x) = 1] \cdot f^{(D)}[g_1(x), g_2(x), \cdots, g_v(x)]}{f[g_1(x), g_2(x), \cdots, g_v(x)]}$$

$$= \frac{p_0(x) \cdot f^{(D)}[g(x)]}{f[g(x)]} \tag{5}$$

If $g_i(x)$ $(i = 1,2,\cdots,v)$ are discrete, inequality $|G_i(x) - g_i(x)| \le \varepsilon_i$ can be written as $G_i(x) = g_i(x)$ for suitable small ε_i. Thus, one can get

$$p_A^*(x) = \frac{p_0(x) \cdot P[G(x) = g(x) \mid D(x) = 1]}{P[G(x) = g(x)]} \tag{6}$$

If sampling procedure is representative (with no error), in general, the density value of denominator in equation (5) and the probability value of denominator in equation (6) can be approximately thought of as constants, say, C^{-1} and C_1^{-1}, respectively. In fact, it is naive because no reason can be used to interpret that the probabilities of $\{G_i(x) = g_i(x)\}$ occurring at different points x are evidently varied. Therefore, identification probabilities in equations (5) and (6) can be approximately written as

$$p_A(x) = C \cdot p_0(x) \cdot f^{(D)}[g(x)] \tag{7}$$

$$p_A^*(x) = C_1 \cdot p_0(x) \cdot P[G(x) = g(x) \mid D(x) = 1] \tag{8}$$

If $G_i(x)$ $(i = 1,2,\cdots,v)$ are independent of each other, equations (7) and (8) can be written as

$$p_A(x) = C \cdot p_0(x) \cdot \prod_{i=1}^{v} f_i^{(D)}[g_i(x)]$$

$$p_A^*(x) = C_1 \cdot p_0(x) \cdot \prod_{i=1}^{v} P[G_i(x) = g_i(x)| D(x) = 1]$$

where $f_i^{(D)}(y_i)$ are the marginal densities of $f^{(D)}(y_1, y_2, \cdots, y_v)$.

Local Favorability of Sampling

It is notable that the objective is to select a set of additional drilling locations in order to determine the pool limits. So, the most favorable locations for additional wells should be located near to the oil-bearing boundary. This subsection presents the probability that a point within the area of interest is located on the curve of the oil-bearing boundary.

Suppose that the extent of the area V of interest corresponds approximately to an oil-bearing trap. In detail, the favorable area V is obtained by regional exploration and geological interpretation; oil-forming, migration, and preservation environments within the area are very good, and can be ignored here. Structural and reservoir environments should be mainly considered.

Assume that A is a set of the points on the curve of the oil-bearing boundary in V. $g_S(x)$ and $g_R(x)$ are the structural and reservoir environment variables, respectively, that is, $g_S(x)$ represents the integrative structure conditions, and $g_R(x)$ represents the integrative reservoir conditions. $G_S(x)$ and $G_R(x)$ are the random functions associated. Set

$$G(x) = [G_S(x), G_R(x)], \quad \text{and} \quad g(x) = [g_S(x), g_R(x)]$$

Hereafter, $G(x)$ is assumed to be continuous. Similar demonstration procedures are also suitable to the situation where $G(x)$ is discrete. From equation (7), continuing to use the other notations above, one can get

$$\begin{aligned} p_A(x) &= C \cdot p_0(x) \cdot f^{(D)}[g(x)] \\ &= C \cdot p_0(x) \cdot f^{(D)}[g_S(x), g_R(x)] \end{aligned} \tag{9}$$

Set $f_S^{(D)}[g_S(x)]$ and $f_R^{(D)}[g_R(x)]$ are the marginal densities of $f^{(D)}[g_S(x), g_R(x)]$.

As discussed in the introduction, the purpose is to design the optimal locations of the appraisal drilling for estimating pool limits. Equation (9) indicates the local favorability of designing an appraisal well location at point x, in which it is important to provide an appropriate model for calculating bivariate probability density $f^{(D)}[g_S(x), g_R(x)]$ that presents the probability of the point x being located on the curve of the oil-bearing boundary.

Linear Hypothesis

Assume that $z_{S_i}(x)$ $(i = 1, 2, \cdots, m_1)$ are the existing structure variables (represent the structural characteristics of the reservoir top surface), and $z_{R_i}(x)$ $(i = 1, 2, \cdots, m_2)$ the existing reservoir variables (represent the geophysical properties of the reservoir). $Z_{S_i}(x)$ $(i = 1,2,\cdots,m_1)$ and $Z_{R_i}(x)$ $(i = 1,2,\cdots,m_2)$ are the random functions associated.

For simplicity, it is naive to assume that

$$f_S^{(D)}[g_S(x)] = \sum_{i=1}^{m_1} \lambda_i f_{S_{T_i}}^{(D)}[z_{S_i}(x)],$$

$$f_R^{(D)}[g_R(x)] = \sum_{i=1}^{m_2} \beta_i f_{Z_{R_i}}^{(D)}[z_{R_i}(x)] \tag{10}$$

where $f_Z^{(D)}[z(x)]$ is the conditional density of $\{Z(x)=z(x)\}$ if $\{D(x)=1\}$ occurs, weights λ_i and β_i are determined by geologist, and $\sum_{i=1}^{m_1} \lambda_i = 1$, $\sum_{i=1}^{m_2} \beta_i = 1$. The differences among λ_i $(i = 1,2,\cdots,m_1)$ or β_i $(i = 1,2,\cdots,m_2)$ reflect the differences of both the credibility of these variables (because of the existence of errors in the measure and explanation for these variables) and the importance of these variables as hydrocarbon indicators. The hypothesis in equation (10) implies that the conditional probability densities, $f_S^{(D)}[g_S(x)]$ and $f_R^{(D)}[g_R(x)]$, of the integrative structure and reservoir variables can be interpreted as the linear combinations of the conditional probability densities of existing variables associated, respectively.

$g_S(x)$ and $g_R(x)$ are not requisite under the linear hypothesis of equation (10), while the conditional density $f_Z^{(D)}[z(x)]$ is required for each available variable $z(x)$.

Using Bayesian formula, one can get

$$f^{(D)}[g_S(x),g_R(x)] = f_S^{(D)}[g_S(x)] \cdot f_R^{(D)}[g_R(x)|G_S(x) = g_S(x)] \tag{11}$$

or

$$f^{(D)}[g_S(x),g_R(x)] = f_R^{(D)}[g_R(x)] \cdot f_S^{(D)}[g_S(x)|G_R(x) = g_R(x)] \tag{12}$$

Intuitively, equation (11) may be more reasonable than (12) in application, because, in general, structure variable is more credible than reservoir variable. So, equation (9) can be written as

$$p_A(x) = C \cdot p_0(x) \cdot f_S^{(D)}[g_S(x)] \cdot f_R^{(D)}[g_R(x)|G_S(x) = g_S(x)] \tag{13}$$

Equation (13) corresponds to the results of repeatedly using Bayesian formula to calculate the posterior probability of $\{D(x)=1\}$ under the condition $\{G_S(x) = g_S(x)\}$ and $\{G_R(x) = g_R(x)\}$, respectively.

$f_R^{(D)}[g_R(x)|G_S(x) = g_S(x)]$ in equation (13) can be thought of as an updated density of $f_R^{(D)}[g_R(x)]$ under the condition $\{G_S(x) = g_S(x)\}$. One can not get any relationship between $f_R^{(D)}[g_R(x)|G_S(x) = g_S(x)]$ and two marginal densities, $f_S^{(D)}[g_S(x)]$ and $f_R^{(D)}[g_R(x)]$, because of the complications of the phenomenon of interest. So, linear model can be used, that is, assume that

$$f_R^{(D)}[g_R(x)|G_S(x) = g_S(x)] = (1-\alpha) f_R^{(D)}[g_R(x)] + \alpha f_S^{(D)}[g_S(x)] \tag{14}$$
$$0 \le \alpha \le 1$$

where α can be termed an *updated degree*. Thus, equation (13) can be written as

$$p_A(x) = C \cdot p_0(x) \cdot f_S^{(D)}[g_S(x)] \cdot \{(1-\alpha) f_R^{(D)}[g_R(x)] + \alpha f_S^{(D)}[g_S(x)]\} \tag{15}$$

If $\alpha = 0$, the equation (15) will degenerate into

$$p_A(x) = C \cdot p_0(x) \cdot f_S^{(D)}[g_S(x)] \cdot f_R^{(D)}[g_R(x)] \tag{16}$$

Equation (16) corresponds to the case where $G_S(x)$ is independent of $G_R(x)$.

For structural pool, in exploration stage, usually, the reservoir variables can be thought of independent of the structure variables. So, set $\alpha = 0$. On the other hand, if the reservoir variables are doubtful to some extent, an appropriate α can be used in order to update $f_R^{(D)}[g_R(x)]$.

ENCODING OF VARIABLE

Calculating $f_Z^{(D)}[z(x)]$ from original variable $z(x)$ corresponds to the encoding procedure in information theory. So, $f_Z^{(D)}[z]$ can be also thought of as an *encoding function*. Value $f_Z^{(D)}[z(x)]$ at location x measures the local favorability of designing exploration wells at the given location, from the point of view of the given variable only and for the given drilling purpose. The variable $f_Z^{(D)}[z(x)]$ can be also termed a *characteristic variable* associated with the original variable $z(x)$.

The maximum values of characteristic variable are obtained at the most favorable locations of designing exploratory wells, and the minimum values at the most unfavorable locations. In general, the favorability of designing a well at a point increases along with the increment of the value of characteristic variable. For the given purpose, in detail, the most favorable locations of designing exploratory wells are located near to the oil-bearing boundary, and the most unfavorable locations are located within the definite oil-bearing area and, particularly, within the definite non-oil-bearing areas. The maximum point of encoding function of a variable can be termed the *favorable characteristic value* of the variable.

The distribution features of characteristic variable depend on the model of encoding function. The curve shapes of encoding function can represent the geologist's adventurous psychology and the credibility of the original data. Encoding function can be classified into the following different types according to their shapes and actions.

    (a) *Conservative* model (figure 1a);
    (b) *Average* model (figure 1b);
    (c) *Adventurous* model (figure 1c);
    (d) *Fuzzy* model (figure 1d).

Models (a), (b), and (c) are appropriate for the encoding of the variable with high credibility, while model (d) is appropriate for the variable with low credibility only.

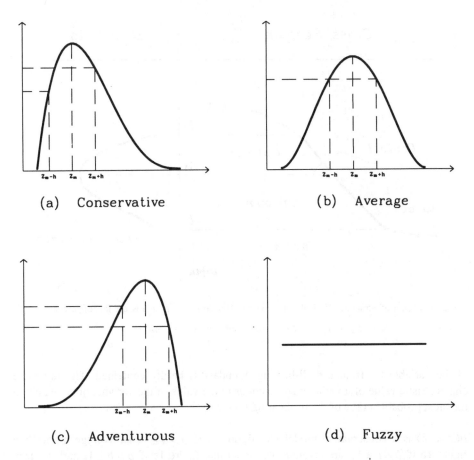

Fig. 1 Encoding Function. Positive axis of abscissa directs the increment of oil-bearing favorability. Ordinate represents encoding value. z_m is the favorable characteristic value of a variable. h is an arbitrary constant.

A simple and reasonable encoding function, which has the features required above, is the following Beta density function:

$$B(z;a,b) = \begin{cases} K(z-t_0)^{a-1}(t_1-z)^{b-1} & z \in (t_0, t_1) \\ 0 & otherwise \end{cases} \qquad (17)$$

where $K = \Gamma(a+b)/[\Gamma(a)\Gamma(b)(t_1-t_0)^{a+b-2}]$, and $\Gamma(\alpha)$ is the Γ–function. The maximum value of $B(z; a, b)$ is obtained at

$$z_m \equiv [(b-1)t_0 + (a-1)t_1]/(a+b-2).$$

(t_0, t_1) can be termed the *favorable characteristic interval* of variable $z=z(x)$. z_m is the favorable characteristic value of the variable.

The favorable characteristic interval of a variable represents such an interval in which the value of the variable near to the oil-bearing boundary is almost contained while the value

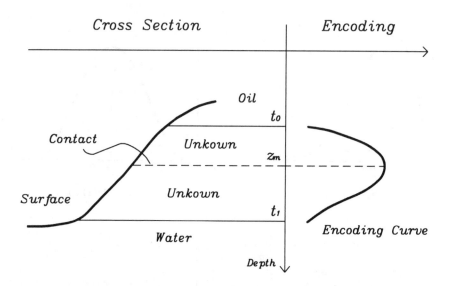

Fig. 2 Encoding of Structural Variable of Reservoir Top Surface. (t_0, t_1) is the favorable characteristic interval of the variable. z_m is the favorable characteristic value of the variable.

of the variable far from the oil-bearing boundary is hardly contained. The favorable characteristic value of a variable corresponds to the value of the variable obtained along the most probable curve of the oil-bearing boundary.

$B(z; a, b)$ is a conservative model like figure 1a if $b > a \geq 1$, an average model like figure 1b if $a = b > 1$, an adventurous model like figure 1c if $a > b \geq 1$, and a fuzzy model like figure 1d if $a = b = 1$. It is clear that $\eta = a / b$ indicates the adventurous degree. So, η can be termed an *adventurous scale*.

The favorable characteristic interval, favorable characteristic value, and adventurous scale should be determined before encoding. Figure 2 shows the encoding of structural variable of reservoir top surface. Conservatism can make that the encoding values above oil-water contact are greater than that below the contact; adventurism is the inverse of conservatism; and averages can use the symmetry encoding curve.

PSEUDO-ENTROPY CRITERION

Using an appropriate criterion or procedure, such as the *moving dividing* procedure [13], based on the conditional probabilities $p_A(x)$ for every location $x \in V$ calculated from equation (15), the optimal locations of the appraisal drilling for estimating pool limits can be selected. This section presents a pseudo-entropy criterion.

Suppose that $\{y_1, y_2, \cdots, y_n\}$ is a sample space for discrete random variable y. $p(y_i)$ denote the occurrence probabilities for realization y_i, in which $\sum\limits_{i=1}^{n} p(y_i) = 1$. Then, the entropy, $H(y)$, of is defined as

$$H(y) = -\sum_{i=1}^{n} p(y_i) \log_r p(y_i) \tag{18}$$

where $r>0$, usually is 2 or e (the base of natural logarithms).

Entropy has be widely used to measure the uncertainty of geological information [4, 5]. It is clear that entropy, $H(y)$, in equation (18) can not be directly thought of as a criterion to select optimally drilling locations.

According to the definition of information entropy, x_i $(i = 1,2,\cdots,n)$ are supposed to be spatial locations. Set $X = \{x_1, x_2, \cdots, x_n\}$. $p_A(x_i)$, calculated from equation (15), are the occurrence probabilities for locations x_i $(i = 1,2,\cdots,n)$. Then, the *pseudo-entropy*, $H_p(X)$, of X is defined as follow

$$H_p(X) = -\sum_{i=1}^{n} p_A(x_i) \log_r p_A(x_i) \tag{19}$$

Pseudo-entropy $H_p(X)$ in equation (19) has the following properties:

(a) $p_A(x_i)$ may not be subject to $\sum\limits_{i=1}^{n} p_A(x_i) = 1$. So, $X = \{x_1, x_2, \cdots, x_n\}$ may not be a sample space for some random variable. Thus, $H_p(X)$ is not an entropy.

(b) Because $f(y) = -y \log_r y$ is strictly incremental if $0 < r < 1$ and $y > e^{-1}$, $H_p(X)$ is strictly incremental if $0 < r < 1$ and $p_A(x_i) > e^{-1}$ $(i = 1,2,\cdots,n)$.

(c) Assume that $0<r<1$. It is clear that $H_p(X) \leq 0$ because of $0 \leq p_A(x_i) \leq 1$ $(i = 1,2,\cdots,n)$. The minimum value, $ne^{-1}/\ln r$, of $H_p(X)$ is obtained at $p_A(x_i) = e^{-1}$ $(i = 1,2,\cdots,n)$. So,

$$H_p^*(X) = -ne^{-1}/\ln r + H_p(X) \tag{20}$$

is positive.

According to the second property of pseudo-entropy, it is reasonable that $H_p(X)$ or $H_p^*(X)$ is thought of as a criterion for optimally selecting drilling locations. Moreover, it is inappropriate if the distance between two drilling locations is too small. Thus, from equation (20), pseudo-entropy criterion can be stated as maximizing the following function:

$$H_p^*(X) = -ne^{-1}/\ln r - \sum_{i=1}^{n} p_A(x_i) \log_r p_A(x_i) \tag{21}$$

$$0 < r < 1, \qquad d(x_i, x_j) \geq L, \qquad d(x_i, x_k^{(0)}) \geq L_0,$$

$$i, j = 1, 2, \cdots, n; \ k = 1, 2, \cdots, m$$

where $d(x_i, x_j)$ is the Euclidean distance between x_i and x_j, $x_k^{(0)}$ $(k = 1, 2, \cdots, m)$ are the existing well locations and, L and L_0 are constants. Because the alternation of parameter r does not have any influence on the results of optimizing drilling locations using pseudo-entropy criterion, r can be set as e^{-1}.

From equation (15), the relative values of identification probabilities can be obtained only. These values may be either too big or small. So, these relative values, in general, are needed to be normalized so that they are not less than 0 and not greater than 1, and most of which obtained at the favorable drilling locations are greater than e^{-1}.

AN EXAMPLE

Available variables, reservoir top surface, porosity, interval velocity, and frequency of seismic wave, which came from a reservoir rock of a gasfield in China, are obtained from Zhao et al. [13]. Figure 3 presents the structural features of the reservoir top surface. The closed bold polygon is the gas-bearing boundary estimated by using 20 borehole data and the geologic interpretation. The boundary can be thought of as the true representation of gas rock distributions in the area. The area of interest is divided into a 26 by 31 grid (25 × 30 cells) as shown in figure 3.

At the exploration stage some years ago, only three exploratory wells were drilled at the points of geologic interest as given in figure 3. The objective is to design four additional exploratory wells on the east of the western fault in order to estimate the pool limits. In this miniature example, reservoir top surface is the structure variable, and the others are reservoir variables. Pseudo-entropy criterion will be used to optimally select the additional drilling locations.

The first step is to calculate the identification probabilities at centers of 750 cells from equation (15). Vague prior is assumed, that is, $p_0(x)$ is a constant. Set updated degree $\alpha = 0.2$ in equation (15). Moreover, based on the distribution features of the existing variables [13], table 1 presents the other parameters of calculating identification probability for this example. Overlapping the structural map of reservoir top surface in figure 3, figure 4 presents the distribution features of identification probabilities (normalized) by using marks.

The centers of 120 cells out of 750 cells, in which the identification probabilities are greater than 0.4, are selected and thought of as candidate locations for the new drilling locations. Set $L = L_0 = 800 \, m$. Using pseudo-entropy criterion, four additional drilling locations are selected and shown in figure 4. It is clear that the optimized drilling locations are near to the gas-bearing boundary. So, the approach discussed here is useful.

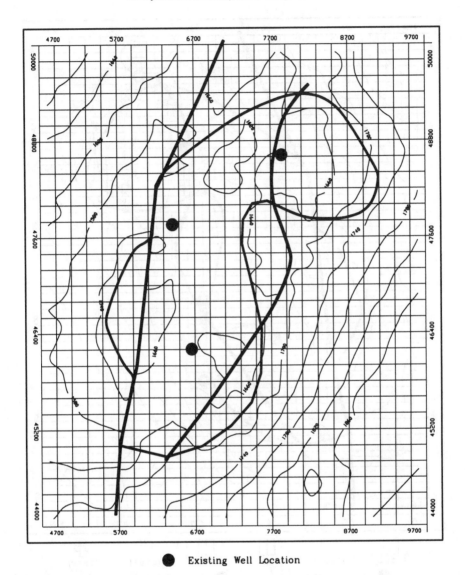

● Existing Well Location

Fig. 3 Structural Map of Reservoir Top Surface. The closed bold polygon is the gas-bearing boundary estimated by using 20 borehole data. The boundary can be thought of as the true representation of gas rock distributions in the area. The area of interest is divided into a 26 by 31 grid.

Table 1. Parameters of Calculating Identification Probability

Variables	Weights	Encoding Parameter a	Encoding Parameter b	Favorable Characteristic Interval
Top Surface	1.0	2.0	2.0	[1640, 1730]
Porosity	0.5	2.0	2.0	[5, 10]
Velocity	0.2	2.0	2.0	[3550, 3750]
Frequency	0.3	2.0	2.0	[32, 38]

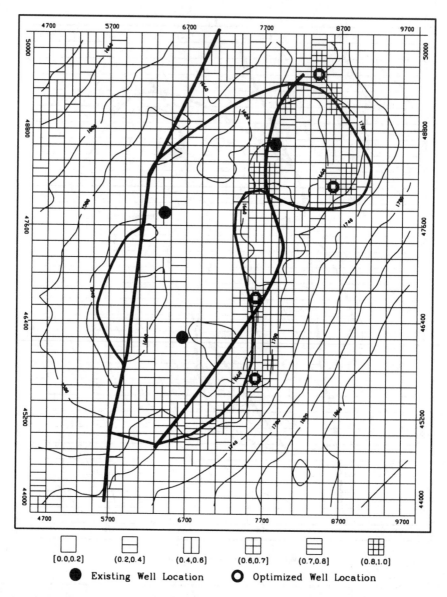

Fig. 4   Optimized Well Locations and Identification Probability Map Overlapping Structural Map of Reservoir Top Surface in Figure 3.

CONCLUSION

It is important to embrace suitable integrating and classifying techniques for the design of drilling locations in geological exploration. Based on the Bayesian law, an approximate model for calculating identification probability is presented. The model can integrate a variety of the existing geophysical variables in geological exploration. An encoding procedure for existing variables is discussed in order to integrate spatial data and calculate identification probability. The encoding procedure can represent the geologist's adventurous psychology and the credibility of the original data.

Entropy has be widely used to measure the uncertainty of geological information. Unfortunately, it can not be directly thought of as a criterion of optimally selecting drilling locations. A pseudo-entropy criterion is proposed in order to design the optimal drilling locations.

Acknowledgments

We thank Dr. Frederik P. Agterberg for reviewing this manuscript and polishing our English.

REFERENCES

1. D. Aspie and R.J. Barnes. Infill-Sampling Design and the Cost of Classification Errors, *Math. geology* 22, 915-932 (1990).
2. C. Barcelo, V. Pawlowsky and E. Grunsky. Classification Problems of Samples of Finite Mixtures of Compositions, *Math. Geology* 27, 129-148 (1995).
3. R.J. Barnes. Sample Design for Geologic Site Characterization. In: *Geostatistics*, v.2. M. Armstrong (Ed.). pp. 809-822. Kluwer, Dordrecht (1989).
4. G. Christakos. A Bayesian /Maximum-Entropy View to the Spatial Estimation Problem, *Math. Geology* 22, 763-777 (1990).
5. G. Christakos. *Random Field Models in Earth Sciences*. Academic Press, Inc. (1992).
6. D.R. Davis, L. Duckstein and R. Krysztofowicz. The Worth of Hydrologic Data for Nonoptimal Decision Making, *Water Resources Research* 15, 1733-1742 (1979).
7. H.Y. Gao, J.H. Wang and P.D. Zhao. The Updated Kriging Variance and Optimal Sample Design, *Math. Geology* 28, 295-313 (1996).
8. U.C. Herzfeld and D.F. Merriam. Optimization Techniques for Integrating Spatial Data, *Math. Geology* 27, 559-588 (1995).
9. M.A. Oliver and R. Webster. A Geostatistical Basis for Spatial Weighting in Multivariate Classification, *Math. Geology* 21, 15-35 (1989).
10. S.K. Thompson. *Sampling*. John Wiley & Sons, Inc., New York (1992).
11. D. Veneziano and P.K. Kitanidis. Sequential Sampling to Contour an Uncertain Function, *Math. Geology* 15, 387-404 (1982).
12. A.G. Watson and R.J. Barnes. Infill Sampling Criteria to Locate Extremes, *Math. Geology* 27, 589-608 (1995).
13. P.D. Zhao, J.H. Wang and H.Y. Gao. Optimization of Drilling Locations and Spatial Sampling Probability. *30th int. Geol. Congress*, 4-14 Aug. 1996, Beijing, China.
14. H. Zhu and A.G. Journel. Formatting and Integrating Soft Data: Stochastic Imaging Via the Markov-Bayes Algorithm. Presented in: *4th Int. Geostatistics Congress*, 13-18 Sept. 1992, Preprints, 12p.

Proc. 30th Int'l Geol. Congr., *Vol.25*, pp. 89-92
Zhao Peng-Da *et al* (Eds)
© VSP 1997

Orderly Rule of Spatial Distribution of Mineralization and Location Prediction to Orebody

WEI MIN   ZHAO PENGDA   SUN JIANHE
Faculty of Earth Resources, China University of Geosciences, Wuhan 430074

Abstract

This paper systematically discusses the universalism of the orderly rule of mineralization   space distribution, expression forms, formation   mechanism, and   gives   examples for their analysis. At the same time,   the sense for prediction   and   prospecting   of   mineralization   orderly distribution is stressed, principles and   methods for the location prediction of orebodies are proposed.

Keywords: mineralization space distribution, orderly rule, location prediction of orebody, wave field model of mineralization, geometry of orebody

Time and space evolution of mineralization in Earth's crust is very complicated. However, its spatial distribution is orderly. It is just the most important task for exploration geologists to study the orderly rule. The rule exists widely and its main expression form are as the follows: zoning nature, equidistant nature, mineralizing wave field, lateral overlapping of ore bodies, hosting regularity of ore pillar, orderly nature of overall configuration et al. The genetic mechanism of orderly distribution concern the following aspects: staying wave principle and self-similarity, the equilibrium principle of total mineralizing materials, host structural   trap and tectonic superposition and deformation, geometric principle of ore bodies, variaton of physical-chemical conditions, geologic background of mineralization and so on.

It is fundamental principle for location prediction to ore bodies to predict the possible position of unknown ore bodies according the orderly distribution rule of mineralization. The geologic anomaly resulting in mineralization is the direct basis for location prediction.

It is the basis way to summarizing orderly distributing rule and location prediction of ore bodies to ascertain their mathematical characteristics (statistical, structure, spatial and geometric) and study the quantity regularity of mineralizing evolution.

The predicting methods applied include wave field model. geometric features of ore bodies, equidistance prediction, self-similarity, trend analysis and mineralization zoning.

The prediction of concealed ore bodies in Kalatongke Cu-Ni ore area in Xinjiang is one of samples of the location prediction methods.

(1)Statstical characteristics of Nickel grade. Two different mineralization types were discovered by screening to the mixed population, the low-value population represents liquating type of magma(lean mineralization)which holds 90% ,the high-value represents injected type of deep ore magma(rich mineralization)which holds 10%, the grade dividing point for these population is 3.0%(Fig.1).

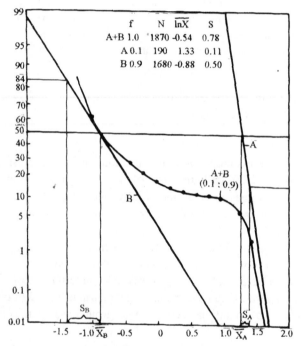

Figure 1. The screening to mixed population of Nickel grade
A--injected component population
B--liquation component population
A+B--mixed population.

(2)Spatial characteristics of ore bodies. The trend analysis to mineralization intensity revealed that rich ore bodies are steep and show discordant intrusion in lean ore bodies(Fig.2).

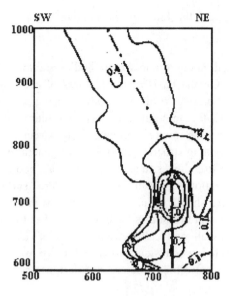

Figure 2. Nickel grade contour map in 28-section
　　　　　　　– · –　ore-controlling structure to lean mineralization
　　　　　　　――――　ore-controlling structure to rich mineralization.

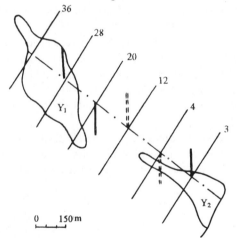

Figure 3. Occuring ideal model for the injected orebodies
　　　　　　　– · – major ore-controlling structure centerline
　　　　　　　―――― known injected orebodies centerline
　　　　　　　==== predicted injected orebodies centerline
　　　　　　　Y1,Y2 olivine gabbro rock bodies
　　　　　　　12 — exploration section and number.

(3)Geometric features of ore bodies. The centerline of ore bodies with different level trends snakely, which was caused by the joint ore-controlling of NW and NNW-trending structure, and the latter is the impounding structure of rich ore bodies of injected type.

(4)Ideal mineralized model of injected ore bodies and location prediction of concealed ore bodies(Fig.3). On the basis of the features of ore-controlling structure and the mathematical characteristics of ore bodies, using geometric method, the ideal mineralization model that mineralization occurred intervally (in heading wall and foot wall)along NW-trending major ore-controlling structure is proposed. The model reveals that NW-trending structure controlled the general occurrence of rock body and the deposits, NNW-trending structure and rich ore bodies distribute wavily by the sides of mineralization centerline(major fracture surface). We predicted that concealed ore bodies should occur in the heading wall of PL.12 and the foot wall of PL.4 because some rich ore bodies have discovered in the heading wall of PL.28 and PL.3 and the foot wall of PL.20, gravity and electrical survey demon strated the existence of integrated geophysical anomaly. Finally, we located concealed bodies by streoplotting to ore-controlling structure, and designed confirmation holes.

REFERENCES

1.Zhao Pengda, Wei Min, Jin Youyu. Statistical analysis in geological exploration. Wuhan. China University of Geosciences Press, 1990.44-45.

2.Zhao Pengda, Zhou Youwu, Wei Min. Geomathematical research on Au mineralization variation. Wuhan. China University of Geosciences Press, 1992.42-57

3.Wei Min. Mathematical characteristics of orebodies and concealed orebody prediction in a mining area in Xinjiang. Chemical Abstracts, 1991.114(20):223

4.Wei Min. Variation rule and variation model for gold mineralization. Chemical Abstracts. 1992,116(20):214

5.Liu shinian. Metallogenic prognosis. Changsha. Zhongnan Industrial University Press, 1993.209

6.Wei Min. Research on geological anomaly and deposit statistical prediction in a mineralization zone in Mianluoning region, Shanxi. China Mathematical Geology, 1995(6).68-85

Proc. 30th Int'l Geol. Congr., Vol.25, pp. 93-102
Zhao Peng-Da *et al* (Eds)
© VSP 1997

Three–Dimensional Simulation of Geologic Structures in Yakumo Geothermal Field, Southwest Hokkaido, Japan

Wei Qiang, Suzuki,Y., Kawakami,N., Takasugi,S.(Geothermal Energy Research and Development Co.,Ltd.) and Kodama,K.(Geological Survey of Japan)

INTRODUCTION

The fractures play important roles for the transport of liquid and vapour. Many fractures have taken place in the Green Tuff region along the Japan Sea side of the Japanese islands since early Miocene, where many Quaternary volcanoes are distributed(Fig.1). The geothermal energy is stored in the deep formations in the Green Tuff region, which has been explored for the power station.

As the structure in the deeper parts is much different from that in the shallower ones showing relatively gentle structure, the conventional method of extrapolating the shallow geologic structure to the deep ones is not applicable.

In order to investigate the deep structures, we tried to apply the Virtual Basement Displacement method (Kodama et al.,1985) to the problem.

OUTLINE OF GEOLOGY

Yakumo geothermal field is studied, which is located in the northern part of the Oshima Peninsula, southwest Hokkaido, Japan(Fig.1). Geologically the area is situated in the Green Tuff region extending along the Japan Sea side of the Japanese islands where energetic volcanic activity has taken place since early Miocene and many active volcanoes and hot springs are distributed, and some geothermal power stations are in operation.

The geology of the area is composed of Neogene and Quaternary formations underlain by the faulted and folded basements of Mesozoic and Paleozoic strata inruded by Cretaceous granitic rocks(Fig.1).

The basements are distributed in the west of the area. The Neogene and Quaternary formations are distributed to the east of the basements and extend in north–south or NW–SE direction and dip gently east or northeastin general.

The Neogene and Quaternary formations are divided into the Usubetsu, Hidarimatagawa, Yakumo, Kuromatsunai and Setana Formations in ascending order(Ishida,1981;Kato et al.,1993). Their age, lithofacies and thicknessare shown in Table 1.

The Usubetsu Formation is the lowest strata of the Neogene system deposited unconformably on the basement, and is in fault contact with it in most of the area. It is distributed in the west part of the area and dips homoclinally 20° to 40° toward east.

The Hidarimatagawa Formation lies conformably on the Usubetsu Formation. It is distributed in the west part of the area and dips homoclinally 30° to 60° toward east.

The Yakumo Formation lies conformably on the Hidarimatagawa Formation, and is distributed in the

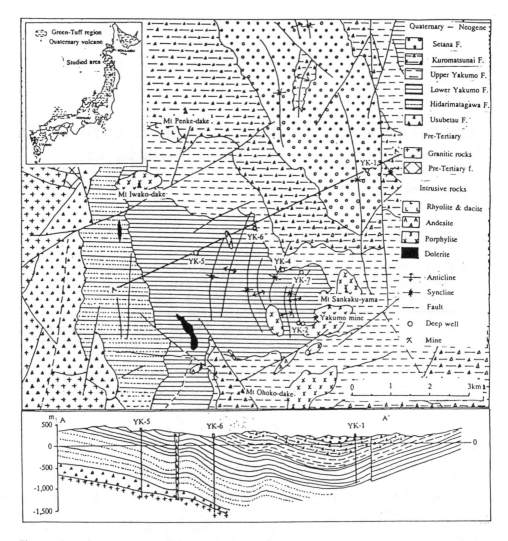

Figure 1. Geologic and index map of Yakumo geothermal field, southwest Hokkaido, Japan.

center of the area. It dips 10° to 50° in general. The lower boundary of the formation run nearly in N–S directionin the west of the area, and the upper one shows NW–SE direction in the northeast and NE–SW direction in the southeast. It is due to the formation of the Namarikawa Uplift Zone in the Yakumo Formation. Referring to the isopachous maps(Fig.2), we can read that the zone was the depocenter in the Lower Yakumo stage and has turned to the uplift zone since the Upper Yakumo stage. Brachy anticlines and synclines in N–S direction are formed in the zone. These complex processes might be resulted in the complex structure in the deep under the zone and its surrounding areas.

The Kuromatsunai Formation is distributed in the center of the area and overlies unconformably on the Yakumo and Hidarimatagawa Formations in the northwest part of the area, but lies conformablly on the Yakumo Formation in the center of the area. It dips homoclinally 20° to 40° toward northeast.

The Setana Formation is distributed in the northeast part of the area, and forms a gentle synclinorium

Table 1. Stratigraphic column

Age	Formation	Lithology	Maximum thickness
Pleistocene	Setana F.	Conglomerate and sandstone / Sandstone	600 m
Pliocene	Kuromatsunai F.	Siltstone, sandstone and volcanic breccia	1200 m
Miocene	Yakumo F.	Mudstone and sandstone intercalating tuff breccia and volcanic breccia	1000 m
		Siltstone and hard shale intercalating tuff	1000 m
	Hidari-matagawa F.	Tuff breccia	800 m
		Sandstone and mudstone	
		Conglomerate	
	Usubetsu F.	Tuff breccia and lava of andesite and basalt	800 m
Pre-Tertiary		Paleozoic-Mesozoic formations granitic rocks	

structure extending in NNW–SSE direction.

Those formations except the Usubetsu Frmation are marine deposits.

In the northwest and south central areas, intrusive rocks in relatively large scale are distributed (Fig.1). The rocks in the former area are andesite, rhyolite and porphyrite intruding the Hidarimatagawa, Yakumo and Kuromatsunai Formations. The rocks in the latter area are andesite and porphyrite intruding the Lower and Upper Yakumo Formations. In the southwest area, dolerite in small scale intruding the Hidarimatagawa Formation is distributed, extending in north–south direction.

Seven boreholes more than 1,000m in depth were drilled in the area and three of them reached the granitic rocks of the basement(New Energy and Industrial Technology Development Organization,1990). Based on the surface exposure and those drilling data, the isopachous maps of the Neogene and Quaternary formations are drawn, as shown in Fig.2. Those figures show eastward or northeastward migration of depocenters in general.

VIRTUAL BASEMENT DISPLACEMENT METHOD

The present geologic structure was formed by the cumulative effect of the overlapping movements of the basement in the past, so the Virtual Basement Displacement method (VBD method) was proposed to reconstruct the complex geologic structures by simulating the process of its movement(Kodama et al.,1985). The most important parameter for VBD method is the surface deformation resulting from the movement of the basement, and the displacement on the surface is assumed to approximate the thickness of the formation during the stage.

The method is laid out in Fig.3. At first, we provide a virtual basement at an arbitral depth and follow the deformation on the surface of the geologic units due to the displacement of the virtual basement by elastic–plastic finite element method under the boundary conditions shown in Fig.4. We compare the calculated deformation of the surface with the isopachous map of the formation, and modify the virtual basement displacement, so that the difference of the two becomes minimal. We repeat the process until the difference becomes sufficiently small, and the VBD at this point will be the optimum basement displacement.

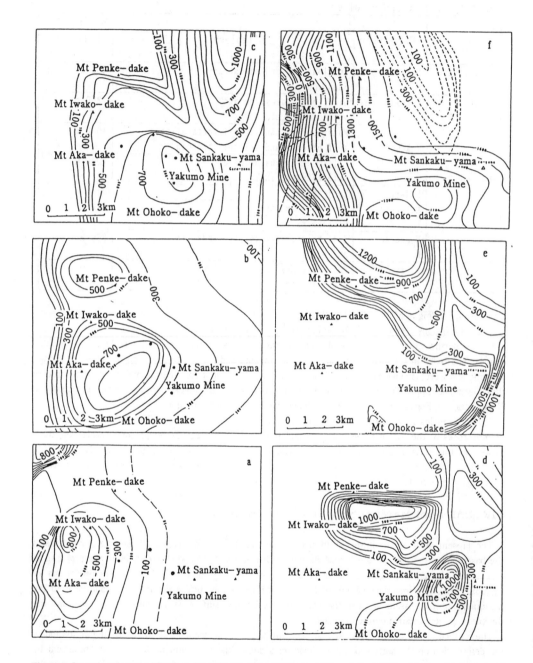

Figure 2. Isopachous maps of Tertiary formations(a–e) and altitudes of the surface of Pre–Tertiary basement and the bottom of Setana Formation(f).

Thus we obtain the incremental deformation of each tectonic stage overthe VBD, and overlap these figures in the order of occurrences up to present. Then we get the model of the present deep structure as the cumulative deformation figures from a certain geologic period.

A geologic unit is supposed to be deformed continuously as a whole, even when faults are partially

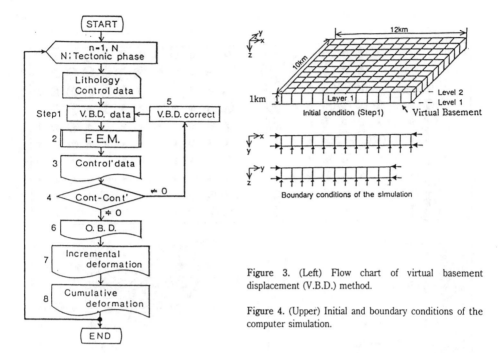

Figure 3. (Left) Flow chart of virtual basement displacement (V.B.D.) method.

Figure 4. (Upper) Initial and boundary conditions of the computer simulation.

formed. The rocks are elastic when the strain is small, but when the strain goes beyond a certain limit, plastic nature arising, decreasing in rigidity. So in the simulation the rocks are supposed to behave as an elastic–plastic body which is not related to strain rate. The transition point of rocks from elastic to plastic nature is called yield point and von Mises' criterion is used for determining the point. Von Mises's criterion means that yield occurs when equivalent stress $\sigma^{\bullet}$ expressed by

$$\sigma^{\bullet} = \{1/2 \ \{ (\sigma_y - \sigma_z)^2 + (\sigma_z - \sigma_x)^2 + (\sigma_x - \sigma_y)^2 + 6(\tau_{yz}^2 + \tau_{zx}^2 + \tau_{xy}^2) \} \ \}^{1/2}$$

exceeds the yield stress σ_y, where σ_x, σ_y, σ_z, τ_{xy}, τ_{yz} and τ_{zx} are components of stress tensor on a point within the geologic unit. The strain at the yield point $\varepsilon_y = \sigma_y / E$ was set approximately 1 percent.

In the simulation, geologic unit is divided into many finite elements of rectangular parallelepiped composed of 6 tetrahedrons. The basement surface which provides displacement boundary condition was assumed to exist 1,000m below the surface of the basement. The displacement were given at 13x11 points which were set every 1,000m on the virtual basement(Fig.4).

The VBD vector is given as vertical and horizontal displacement vector components. In the present simulation, the vertical components were given at first and they were adjusted at each point, so that the difference of the displacement of the surface and the isopachous line became sufficiently small. If the adjustment of the vertical component was not sufficient to decrease the difference, the horizontal component should be adapted, but this was not necessary in the case of the present study. This is explained due to the very small horizontal displacement of the basement, which is composed of rigid Paleozoic and Mesozoic folded and faulted formations intruded by Cretaceous granitic rocks.

RESULT OF SIMULATION

a. Strain, principal stresses and fractures

The experiment of the VBD method gives the distribution of incremental deformation, strain, principal stresses, fractures and cumulative deformations at each stage.

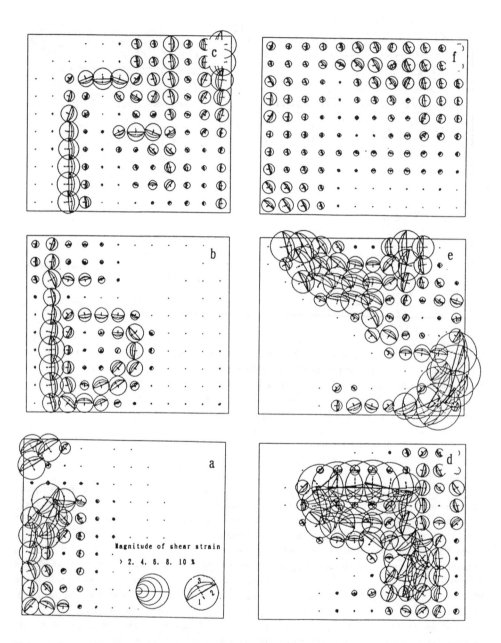

Figure 5. Shear faults in the basement at the end of deposition of each formation. a.Usubetsu Formation, b.Hidarimatagawa Formation, c.Lower Yakumo Formation, d.Upper Yakumo Formation, e.Kuromatsunai Formation, f.Setana Formation.

In the Usubetsu stage, a depocenter was formed in the west, and strain was concentrated along the west side of the depocenter, where north−south trending and easterly stepping down faults were formed(Fig.5 a). Subordinate east−west trending faults were brought about on the north and southsides of the center. They went down toward the center.

In the Hidarimatagawa stage, two depocenters were built in the center of the area, main one of which was situated about 3km eastward compared with that of the former stage(Fig.5 b). Surrounding the depocenters, strain concentration occured, where north–south trending and easterly going down faults took place along the west side of the center, and east–west trending ones occurred along their north and south sides, each of which stepped down southerly and northerly respectively.

In the Lower Yakumo stage, two depocenters were formed in the south central and northeast parts, which were shifted more eastward and north–eastward compared with that of the former ones(Fig.5 c). North–south trending faults occurred along the east and west sides of the center, and east–west trending faults took place along the northwest side of thesouth central depocenter. Each of those faults went down toward the depocenter.

In the Upper Yakumo stage, three depocenters were formed in the north central, southeast and northeast parts(Fig.5 d). In this stage the most intense deformation proceeded and the Namarikawa Uplift Zone which had been subsided before the stage appeared. Surrounding the first depocenter, large strain was accumulated, and east–west trending faults were formed. North–south trending faults were predominantly built in the second depocenter. They went down toward the depocenters.

In the Kuromatsunai stage, three depocenters were built in north central, northeast and southeast parts(Fig.5 e). The intense deformation proceeded and the Namarikawa Uplift Zone kept its movement in this stage. Strain concentration was observed surrounding those depocenters and faults were formed parallel to their boundaries, and stepped down toward the centers.

The last stage is the time when the deformation of geologic units was completed as the present state. The simulation experiment at this stage was operated so as to fit the present form of the Kuromatsunai Formationand the basement. North–south and NNW–SSE trending faults were predominantly formed, accompanying a basin–like down–warping deformation extending in NNW–SSE direction in the east and westward upheaval movement in the west(Fig.5 f).

b. Three dimensional figure

The geologic structure of formations has been formed through the process of accumulation of their own and upper formations due to the basement deformation, so the geologic structure of lower ones is more complex than that of shallower ones. The figure of the base of each formation at the last stage by the V.B.D. method is shown in Fig. 6 .

The figure of the base of the Setana Formation shows a simple synclinal structure extending in NNW–SSE direction in the east of the area(Fig. 6 g).

The bottom figure of the Kuromatsunai Formation exhibits a more steep synclinal structure running in NNW–SSE direction in the northeast part and an anticlinal structure extending in ENE–WSW direction in the southeast part(Fig. 6 f).

The figure of the base of the Upper Yakumo Formation shows an east–west trending anticline and syncline in the north central part, and a north–south trending syncline and a NE–SW trending anticline in the southeast part(Fig. 6 e). A syncline is left in the north central part.

The bottom figure of the Lower Yakumo Formation is similar to that of the Lower Yakumo Formation, but a little steeper in inclination(Fig. 6 d).

The base figure of the Hidarimatagawa Formation is also similar to those of the Upper and Lower Yakumo Formations but more undulated(Fig. 6 c). An anticline appeared in the south central part.

The figure of the bottom of the Usubetsu Formation is comparable with that of the Hidarimatagawa Formation. The figure of the virtual basement is also similar to those figures(Fig. 6 b).

Figure 6a shows the deformation of the virtual basement.

Figure 7 exhibits the bird's eye view of the base figure of each formation corresponding to Figure 6.

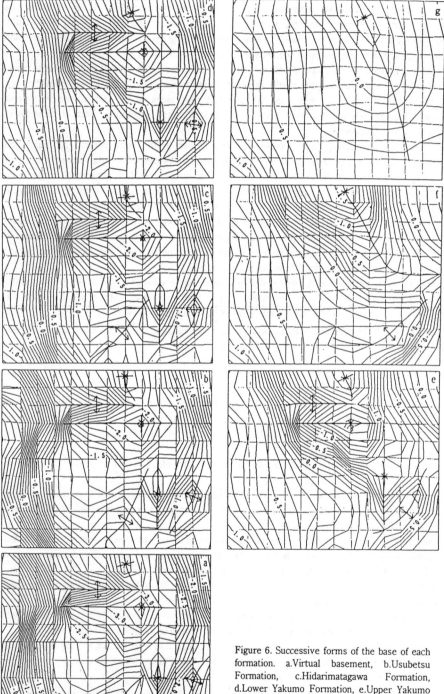

Figure 6. Successive forms of the base of each formation. a.Virtual basement, b.Usubetsu Formation, c.Hidarimatagawa Formation, d.Lower Yakumo Formation, e.Upper Yakumo Formation, f.Kuromatsunai Formation, g.Setana Formation.

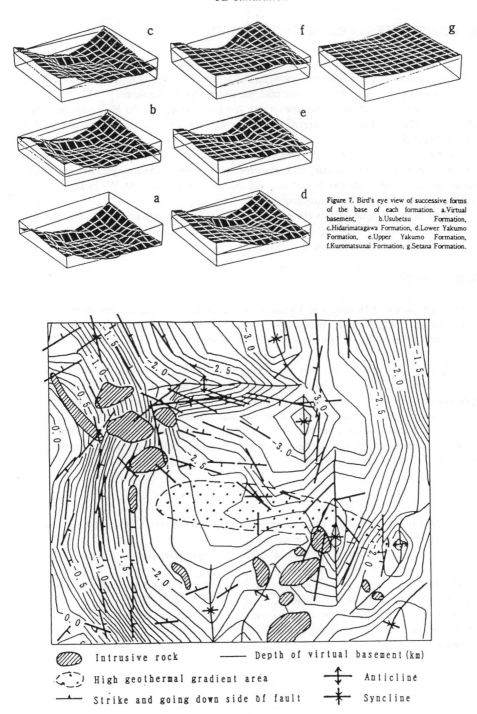

Figure 7. Bird's eye view of successive forms of the base of each formation. a.Virtual basement, b.Usubetsu Formation, c.Hidarimatagawa Formation, d.Lower Yakumo Formation, e.Upper Yakumo Formation, f.Kuromatsunai Formation, g.Setana Formation.

Intrusive rock ——— Depth of virtual basement (km)

High geothermal gradient area ⬍ Anticline

Strike and going down side of fault ✳ Syncline

Figure 8. Deformed figure at the last stage by the virtual basement method and major faults formed at all stages and intrusive rocks.

INTRUSIVE ROCKS IN RELATION TO FAULTS

Fig. 8 exhibits the deformed figure of the virtual basement at the last stage and major faults formed at all stages. It shows that the faults run nearly parallel to the boundaries of each tectonic unit, and they went down toward the center of basin, so the basement displays block—like figure.

We must point out that the Namarikawa Uplift Zone corresponds to the block—like basin of the basement which is characterized by high thermal gradient. Many intrusive rocks of andesite, porphyrite and dolerite are distributed.

On the periphery of the basin where many faults were formed. In the west north—south trending faults were formed at the Usubetsu, Hidarimatagawa and Lower Yakumo stages, and the basement inclines steeply toward east(Fig.8). In the southeast area north—south and NE—SW trending faults were formed at the Upper Yakumo and Kuromatsunai stages.

The High geothermal gradient area observed on the Namarikawa Uplift Zone. It had subsided until the end of the Lower Yakumo stage, and then turned touplift. This fact suggests some endogeneous thermal process activating the tectonic movement.

CONCLUSION

The Virtual Basement Displacement method is applyed to show the geologic development and three dimensional figure of geologic structure of Yakumo geothermal field, and the strain and fault distributions at each stage and the basement figure of each formation at the last stage are brought as shown in Fig.5,6 and 7.

The figures of fault development show that north—south trending faultsare predominant in the west, east—west trending ones in the central area and north—south trending ones in the east. The basement figure exhibits block—like structure bounded by those faults. Intrusive rocks are formedon those faults.

The geothermal resources are expected under the Namarikawa Uplift zoneshowing high gepthermal gradient, but faults are not developed so much, so good hot dry rock reserves must be seached for there.

The three dimensional figure of geologic structure by VBD method were drawn, following the geologic process, so they must be more natural than those by the other methods.

REFERENCES

1. Ishida, M.(1981) Geology of the Yurappu—dake District. Quadrangle series, scale 1:50,000, Geol. Surv. Japan, 64p.(in Japanese with English abstract)
2. Kato,M., Katsui,Y., Kitagawa,Y. and Matsui,M.(1993) Hokkaido district,Geology of Japan 1.337p., Kyoritsu—shuppan Co., Tokyo.(in Japanese)
3. Kodama,K., Long Xue—ming and Suzuki,Y.(1985) Structural analysis of deep—seated volcanic reservoirs by tectonic simulation. US ESCAP, CCOP Tech. Bull., 16, 61—79.
4. New Energy and Industrial Technology Development Organization (1990) Geothermal development promotion survey at Yakumo district, Hokkaido. Rep.Geotherm. Devel. Prom. Surv., no.19,1163p.(in Japanese)

Proc. 30th Int'l Geol. Congr., Vol.25, pp. 103-114
Zhao Peng-Da *et al* (Ed.)
© VSP 1997

Three Dimensional Mathematical Models of Geological Bodies and Their Graphical Display *

ZHANG JUMING, LIU CHENGZOU, SUN HUIWEN

(Institute of Geology, Chinese Academy of Sciences, Beijing 100029, China)

Abstract

Various geological information may be extracted from data of a given geological body. They can be measured in field or obtained in laboratories. Usually geological data are spatially scattered and discontinuous. Therefore, for treating geological data and simulating various geological bodies, mathematical fitting functions are necessary. In this paper several types of fitting functions have been established. For the purpose of realizing three dimensional mathematical simulation of geological bodies and their graphical display, special programs have been compiled. Using these programs regularities in distribution of geological data and spatial characteristics of geological bodies can be illustrated. Three dimensional mathematical models of geological bodies and their graphical display are powerful tools for geological studies and they can raise efficiency of geological works obviously.

Keywrods: 3-D mathematical models, Fitting function, Graphical display

INTRODUCTION

Geological information can be obtained from various types of geological data, such as: topographic inequality, ground water table, faults, joints, distributions of loose stuff, stratigraphic layers, geophysical and geochemical data. Geologists can obtain these data in fields or in laboratories. Usually, geologists process the above-mentioned information by mathematical methods before utilizing them. In this paper mathematical fitting functions are established and they are used for processing geological data and 3-D simulation of various geological bodies. Several case studies are given. The research results show that the 3-D modelling of gological bodies by fitting functions is useful and powerful method in geology.

3-D MATHEMATICAL MODELS OF GEOLOGICAL BODIES

Using different fitting functions, models of geological bodies can be established for various types of geological data. Geological data can be classified into following two types: plane

---

* This paper is partial result of research subject (No. 49572164) aided by National Natural Science Foundation of China.

data in space and curved surface data in space. Every type of data has its own fitting function, which will be introduced as follows.

Fitting function for simulating plane data in space
Faults and joints in rock bodies may be treated approximately as plane in space. A fault plane can be fixed in space if the following location parameters are determined. These are:
1) location point on plane Pc (x_c, y_c, z_c);
2) dip direction α and dip angle β;
3) fault extension along the dip direction s.
Using these parameters the normal vector to the fault(Fig. 1), $\bar{n}$, can be expressed as follows:

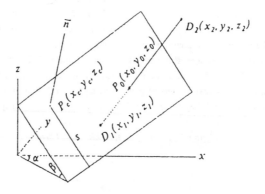

Fig.1   Sketch of space plane

$$\bar{n} = n_x \vec{i} + n_y \vec{j} + n_z \vec{k} \qquad (1)$$

Where $\vec{i}, \vec{j}, \vec{k}$ are the unit vectors of x, y and z axises.

$$\left. \begin{aligned} n_x &= \sin\beta \cdot \cos\alpha \\ n_y &= \sin\beta \cdot \sin\alpha \\ n_z &= \cos\beta \end{aligned} \right\} \qquad (2)$$

If one point P (x, y, z) is on the fault,the line $P P_c$ is perpendicular to $\bar{n}$, so that the relationship equation of x, y and z can be written as:

$$(x - x_c)n_x + (y - y_c)n_y + (z - z_c)n_z = 0 \qquad (3)$$

It is usually needed in models to judge the relation of two points in space with a given fault . For example, whether the two points are located at different side or at same side of the fault. If they are at different side, where is the intersection of the two points linking line with fault? The following equation offers a judge method. Suppose that there are two points $D_1(x_1, y_1, z_1)$, $D_2(x_2, y_2, z_2)$ and a line passing through D_1 and D_2. See Fig.1,then the intersecting point $P_0(x_0, y_0, z_0)$ of the line with fault can be calculated by

$$
\left.\begin{array}{l}
x_0 = x_1 + (x_2 - x_1) \cdot u \\
y_0 = y_1 + (y_2 - y_1) \cdot u \\
z_0 = z_1 + (z_2 - z_1) \cdot u
\end{array}\right\}
\tag{4}
$$

Where

$$
u = [(x_c - x_1) \cdot n_x + (y_c - y_1) \cdot n_y + (z_c - z_1) \cdot n_z] / [(x_2 - x_1) \cdot n_x + (y_2
$$
$$
- y_1) \cdot n_y + (z_2 - z_1) \cdot n_z]
\tag{5}
$$

Here u is a length from D_1, parallel to $D_1 D_2$ and measured by using $D_1 D_2$ as length unit. If u is in the same direction as D_1 D_2 in, it is positive, otherwise, nagative. Then if $u < 0$ or $u > 1$, D_1 and D_2 are at the same side of the fault, If $0 < u < 1$, D_1, D_2 at different side of the fault, and the linking line of D_1 and D_2 passes through fault at $P_0(x_0, y_0, z_0)$. The x_0, y_0 and z_0 then can be determined by formula (4). If $u = 0$ or $u = 1$, then D_1 or D_2 is just in the fault plane. Finally, if the denominator of u equals zero, the linking line will parallel to the plane. When the height z_0 is below $z_c - s \cdot \sin\beta$, the intersection will be located out of the fault plane and be unreal.

The space plane may be used to describe the distribution of joints as well provided that a set of joins was classified into several groups by statistic method, each has its mean volume density, diameter, dip direction and dip angle and their variances. Supposing that each of the joints has disk like shape, then, by random sampling, these joints planes can be determined in the given geological body with random arrangement.

Fittig functions for simulating curved surface data in space
Surface fitting functions can be used to describe such data as topographic inequalities, ground water table, strata and geophysical & geochemical data distributed in space and so no. Suppose that, there is a point i (x_i, y_i, z_i), in space, at where a some physical datum A_i is recorded. The A_i then has an influence on its surroundings. Generally, the influence value $W(r_i)$ at r_i can be written as follows.

$$
W(r_i) = A_i \left(\frac{r_i^2}{R^2} \ln \frac{r_i^2}{R^2} + 1 - \frac{r_i^2}{R^2} \right)
\tag{6}
$$

Where $r_i^2 = (x - x_i)^2 + (y - y_i)^2 + (z - z_i)^2$, R the influence radius.

The influence value $W(r_i)$, which decreases as r_i increases, ranges from A_i to zero and when $r_i = 0$ or $r_i = R$ the $dW(r_i) / dr_i = 0$.

1) Fitting function for simulating mono-surface data
A mono-surface fitting function in this paper means that for any independent point P (x, y) in base plane(plane $z = 0$), the function has only one dependent value which represents the fitting surface's height. Such function can imitate the existing topography, water table, weakened layer and so on. Giving a set of N measurement points x_i, y_i, z_i $(i = 1, 2, \cdots, N)$ which represent the sample data obtained from a some mono-surface. The fitting function then can be made by adding the formula (6) N times with different independent r_i and different coefficient A_i as follows.

$$W_2(x,y) = \sum_{i=1}^{N} A_i \left(\frac{r_i^2}{R^2} \ln \frac{r_i^2}{R^2} + 1 - \frac{r_i^2}{R^2} \right) \tag{7}$$

Where $r_i^2 = (x-x_i)^2+(y-y_i)^2$, $i = 1, 2, \cdots N$.

In order to determine the coefficient A_i, it is necessary to establish a set of N linear equations in N unknowns.

$$Z_i = \sum_{j=1}^{N} A_j \left(\frac{r_{ij}^2}{R^2} \ln \frac{r_{ij}^2}{R^2} + 1 - \frac{r_{ij}^2}{R^2} \right), \quad (i = 1,2,\cdots,N) \tag{8}$$

Where $r_{ij}^2 = (x_i-x_j)^2+(y_i-y_j)^2$. Thus, the equation (7) describes a space surface witch passes through all the sample data and is continuouse and smoothing everywhere.

2) Fitting function for simulating multi-surface data
A multi-surface fitting function in this paper means that for any independent point $P(x, y, z)$, the function has a correspoinding dependent value. Inversely, if the dependent value was given as a constant, then the function represents an iso-value surface in space.
Such function can imitate any geophysical or geochemical data. Supporse that a set of scattered data are $x_i, y_i, z_i, u_i (i = 1,2,\cdots N)$, where x_i, y_i, z_i are coordinats, and u_i represents some physical quantity recorded at (x_i, y_i, z_i), then as eq.(7) and (8), the function and the set of N linear equations to solve A_i can be written as follows.

$$W_3(x,y,z) = \sum_{i=1}^{N} A_i \left(\frac{r_i^2}{R^2} \ln \frac{r_i^2}{R^2} + 1 - \frac{r_i^2}{R^2} \right) \tag{9}$$

$$u_i = \sum_{j=1}^{N} A_j \left(\frac{r_{ij}^2}{R^2} \ln \frac{r_{ij}^2}{R^2} + 1 - \frac{r_{ij}^2}{R^2} \right), (i = 1,2,\cdots,N) \tag{10}$$

Where $r_i^2 = (x-x_i)^2+(y-y_i)^2+(z-z_i)^2$
$r_{ij}^2 = (x_i-x_j)^2+(y_i-y_j)^2+(z_i-z_j)^2$
A_i, the N unknowns, will be determined by solving eq.(10).Under the condition of eq.(10), the $W_3(x, y, z)$ will pass through all the measured data and is continuous everywhere. So that the function can represents the distribution of this kind of data in geological body properly.

3) Fitting function for simulating stratum data
Strata are arranged in sequence, and each stratum has its own thickness. The eqs.(9) and (10) can also be used to simulate the stratum data providing that some changes are made in eq.(10). It is needed to label the stratum sequence from up to down. If we have L surfaces(the boundary surfaces of layer from $K=1$ to L), then consequently we have $L-1$ strata, see fig. 2.

The layer's dependent variable can be written as

$$V_i = \begin{cases} 1 \\ \sum_{j=1}^{i-1} V_j H_j / H_{max} + 1 \end{cases} \quad \text{when } i \begin{cases} = 1 \\ \neq 1 \end{cases} \tag{11}$$

In order to fit all the layers, a set of measured data is needed which can be represented as x_i, y_i, z_i and K $(i=1, 2, \cdots, N)$. where x_i, y_i, z_i are layer's location coordinates, K is the sequence number of the layer surface on which the (x_i, y_i, z_i) is. Then the eq. (10) can be rewritten as

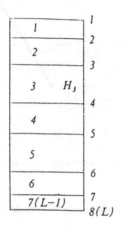

Fig.2 Strata sequence

$$V_k = \sum_{j=1}^{N} A_j \left(\frac{r_{ij}^2}{R^2} \ln \frac{r_{ij}^2}{R^2} + 1 - \frac{r_{ij}^2}{R^2} \right), (i = 1, 2, \cdots, N) \tag{12}$$

The coefficients $A_i (i=1, 2, \cdots N)$ then can be solved by using eq. (12). Thus, inversely, if given $W_j(x, y, z) = V_k (K=1, 2, \cdots, L)$, the eq. (10) describes all the L layer surfaces distributed in the geological body.

The strata may be discontinuous when they pass through a fault. So it is necessary for the eq.(9) to possess a property of local discontinuity (Fig.3).

It can be realized by rewriting r_i and r_{ij} in eqs.(9) and (12) as belows.

Considering points $P(x, y, z)$, $P_i(x_i, y_i, z_i)$ and $P_j(z_j, y_j, z_j)$, if P and P_i are at different side of a fault judging from eq.(5) and the intersection $P_0(x_0, y_0, z_0)$ from eq.(4) is within the fault plane, then r_i will be rewritten as

$$r_i = [(x - x_i)^2 + (y - y_i)^2 + (z - z_i)^2]^{1/2} + \{[(x - x_b)^2 + (y - y_b)^2 + (z - z_b)^2]^{1/2}$$
$$+ [(x_i - x_b)^2 + (y_i - y_b)^2 + (z_i - z_b)^2]^{1/2} - [(x - x_i)^2 (y - y_i)^2 + (z$$
$$- z_i)^2]^{1/2}\} \cdot A_k \tag{13}$$

Similarly, if P_i and P_j are at different side of the fault, the r_{ij} will be

$$r_{ij} = [(x_i - x_j)^2 + (y_i - y_j)^2 + (z_i - z_j)^2]^{1/2} + \{[(x_i - x_b)^2 + (y_i - y_b)^2 + (z_i$$
$$- z_b)^2]^{1/2} + [(x_j - x_b)^2 + (y_j - y_b)^2 + (z_j - z_b)^2]^{1/2} - [(x_i - x_j)^2 + (y_i$$
$$- y_j)^2 + (z_i - z_j)^2]^{1/2}\} \cdot A_k \tag{14}$$

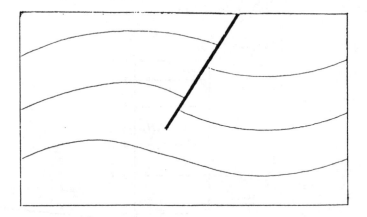

Fig.3 discontinuity of the strata function

where $z_b = z_c - s \cdot \sin\beta$

$\quad x_b = x_c - (z_c - z_b) \cdot ctg\beta \cdot \cos\alpha$

$\quad y_b = y_c - (z_c - z_b) \cdot ctg\beta \cdot \sin\alpha \quad$ (See fig.1.)

A_k is a fault coefficient which controls the faulted distance of the layer through fault.
Obviously, by substituting eq.(13) and (14) into eq.(9) and(12), the layers will have discontinuities when they pass through the fault.

GRAPHICAL DISPLAY OF THE MODELS

A graphical display program has been compiled for the models to provide a means of displaying the function fitting results.

Profile display
Any profile in the geological body can be set up if its location parameters are given. These parameters are(see fig.4)

1) $p_1(x_1, y_1)$ and $P_2(x_2, y_2)$ --the start and end points of the profile in the base plane.
2) z_1--the profile bottom height.
3) V_0--the profile width.
4) β--the dip angle of the profile.

Using these parameters, the profile length u_0, profile top height z_2 and the angle between x axis and the line p_1p_2 can all be determined.

$$u_0 = [(x_2 - x_1)^2 + (y_2 - y_1)^2]^{1/2}$$

$$z_2 = z_1 + v_0 \cdot \sin\beta$$

$$\alpha = \begin{cases} \pi/2 \\ -\pi/2 \\ \text{arctg}[(y_2 - y_1)/(x_2 - x_1)] \\ -\text{arctg}[(y_2 - y_1)/(x_2 - x_1)] \end{cases} \quad \text{When} \begin{cases} x_1 = x_2, y_1 > y_2 \\ x_1 = x_2, y_1 < y_2 \\ x_1 < x_2 \\ x_1 > x_2 \end{cases} \quad (15)$$

Thus, if any point $P(u, v)$ in profile with u_0, v_0 as its local coordinates, the $P(x, y, z)$ in 3-D space can be written as

$$x = x_1 + u \cdot \cos\alpha - v\cos\beta \cdot \sin\alpha$$
$$y = y_1 + v \cdot \cos\alpha \cdot \cos\beta + u \cdot \sin\alpha$$
$$z = z_1 + v \cdot \sin\beta \qquad (16)$$

Because any fitting function independant values are represented in space by x, y or x, y, z, it is necessary to translate u, v into x, y or x, y, z in order to calculate the function values at that point u, v.

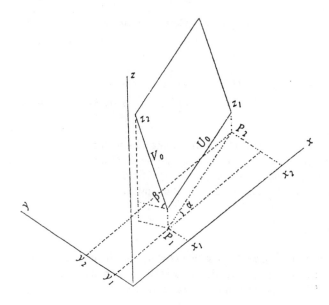

Fig.4 Profile location parameters

Stereoscopic display of the models
Any block in the geological body can also be determined by giving the block location parameters(see fig.5). These are:
1)x_0, y_0, z_0--the block origial at its left lower corner.
2)z_1--block top height.
3)Xl, Yl --block length and width.
4)α_0--the angle between x axis and the Xl.

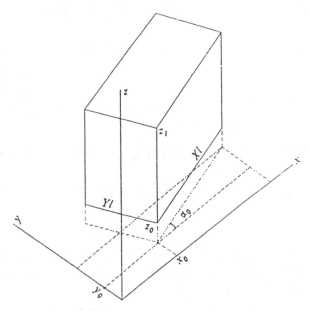

Fig.5 block location parameters

The following projection equation is needed to make the figure have stereoscopic sense.

$$
\left.
\begin{aligned}
u &= x(\cos^2\alpha + \sin^2\alpha\cos^2\beta)^{1/2}\cos\alpha_x - y(\sin^2\alpha + \cos^2\alpha\cos^2\beta)^{1/2}\cos\alpha_y \\
u &= x(\cos^2\alpha + \sin^2\alpha\cos^2\beta)^{1/2}\sin\alpha_x + y(\sin^2\alpha + \cos^2\alpha\cos^2\beta)^{1/2}\sin\alpha_y - z\sin\beta
\end{aligned}
\right\}
\qquad (17)
$$

Where
u -- the abscissa, v -- the ordinate which coincidant with z axis.
α-- the direction angle of vision, β-- the dip angle of the view plane,
$\alpha_x = \mathrm{arctg}(\sin\alpha\cos\beta / \cos\alpha)$, $\alpha_y = \mathrm{arctg}(\cos\alpha\cos\beta / \sin\alpha)$.
With different α and β, the user can view it in any directions.

CASE STUDIES

Supposing a given geological body which contains information of topography, water table, faults, random joints, strata, intrusion massive and so on, a series of case studies have been made and their figure display can be showed as follows:

Case study 1
Fig. 6 is a graphycal disply of result of case study 1. It is a 3–D overall view of a given geological block. It consists of topography, ground water ($\vee$) weakened layer (I), faults and strata. Original coordinates of the block (left lower conner) are $x_0 = 0, y_0 = 0, z_0 = 0$ and
$Xl = 100$, $Yl = 100$

Case study 2
Fig 7 is a graphycal display of result of case study 2. It is a block with random joints and an intrusion massive at the left lower conner.

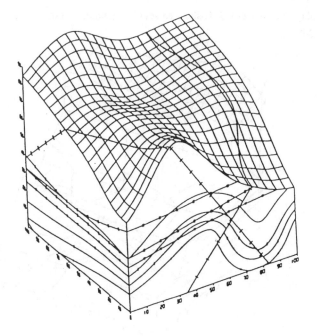

Fig. 6   3–D block 1

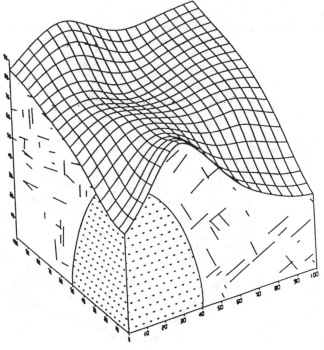

Fig. 7   3–D block 2

Zhang, J. and others

Case study 3

Fig. 8 is a display of case study 3. It shows a set of step shape multi–slopes cut out from the front of the Fig.6

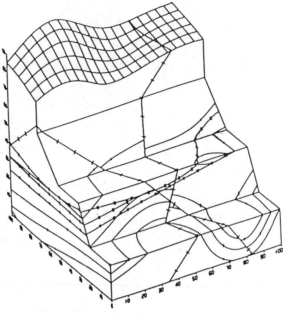

Fig. 8   3–D blick 3

Case study 4

Fig. 9 is a result of case study 4. It shows an additional massive added to the Fig.8

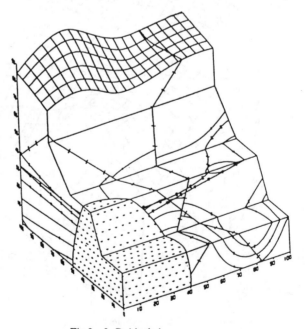

Fig.9   3–D block 4

Case study 5

Fig. 10 is a display of result of case study 5. It shows a block only with iso-lines of some geological data.

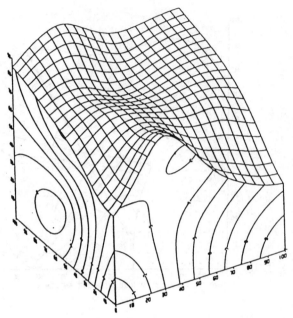

Fig. 10   3−D block 5

Fig.11 is a profile cut from Fig.6 and 7 along its diagonal. The profile start point x1 = 0,y1 = 0, end point x2 = 100,y2 = 100.

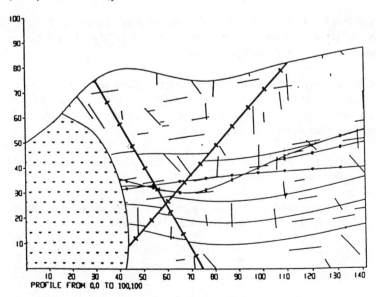

Fig.11   Profile 1

Fig.12 is a profile section cut from Fig.6 and 7. the start point x1 = 0,y1 = 0, end point
x2 = 100, y2 = 0,with dip angle = 60 ° .

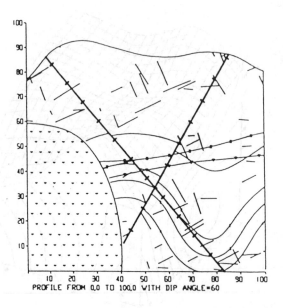

Fig. 12   Profile 2

CONCLUSIONS

3–D modelling based on establishment of fitting functions is useful and powerful method in
geology. It can be used for treating various kinds of geological data and realizing 3–D
mathematical simulation of geological bodies.

Geological data may be divided into two main types: plane data in space and curved surface
data in space. Furthermore, the curved surface data in space can be divided into mono–surface data, multi–surface data and stratum data. For every type of geolocial data, corresponding fitting function has been established.

Method of graphycal disply of 3–D models are discussed in this paper. Method of profile
display and stereoscopic display are introduced.

Finally, 5 case studies and 2 profiles are given. The case studies show that satisfactory results are obtained using the introduced methods in this paper.

Proc. 30th Int'l Geol. Congr., Vol.25, pp. 115-124
Zhao Peng-Da *et al* (Ed.)
© VSP 1997

The Simulation of Braided Channels in Two Dimensions with Random Walk Model

WANG JIA-HUA

Department of Computer Science, Xi'an Petroleum Institute, Xi'an 710065, P. R. China

ZHANG TUAN-FENG

Department of basic Sciences, Xi'an Petroleum Institute, Xi'an 710065, P. R. China

HUANG CANG-DIAN

Department of Computer Science, Xi'an Petroleum Institute, Xi'an 710065, P. R. China

Abstract

In this paper, random walk model is used to simulate 2D distribution of braided channels with frequent connections and tributaries. 2D gridding data   pp(i, j), as a linear combination of three parameters: permeability, porosity, and shale percentage, is used to discriminate the grid node: channel, or shale, or sand between channels and shales. Fractal Brownian Motion is used to simulate 2D distribution of these   three parameters in wells. Channel center lines are defined at first, then channel boundaries are considered. At the end of the paper, a case study is made in the area Sheng-84, with 100 wells, located in Liaohe Oil Field in China.

Keywords: random walk, simulation, realization, braided channel

INTRODUCTION

The sediment in the study area comes from the north east direction, is located in the sub-area between deltaic front fan and deltaic plain, and consists of braided channels and sand between channels and shale. Because of the weak water dynamic capacity during the sedimentary process of deltaic fan, sands between channels and point bars are few, and braided channels are the main skeleton in the study area.

The individual channel sand body in the reservoir has the shape of belts and lenses. All the braided channels spread out along the direction from the north east to the south west. The channel width is small, and the branching of channel happen frequently during the sedimentary process, therefore these braided channels bypass, connect each other, and intersect each other frequently.

Reservoir petrophysical parameters vary greatly .For example, in the same layer, permeability differs from place to place by a factor of ten or more.

CONDITIONAL SIMULATION OF RESERVOIR PETROPHYSICAL PARAMETERS

Because of the strong heterogeneity of the reservoir, factual Browning motion model is used to model the distribution of geophysical parameters.

Geophysical parameters are modelled as a random field $\{Z(x); x \in D\}$ in two dimensions. Suppose the increments of the random field satisfy Gauss process removed from a trend. In the study, the first order trend surface is used. The expectation $EZ(x)$ is chosen as follows:

$$f^T(x)\beta = \beta_1 + \beta_2 x_1 + \beta_3 x_2$$

Here: $\beta^T = (\beta_1, \beta_2, \beta_3)$.

Considering the process without the trend:

$$Y(x) = Z(x) - f^T(x)\beta$$

Here : $Y(x)$ is a stationary process with $EY(x) = 0$. Let D_L be a $(2^n + 1) \times (2^n + 1)$ grid system in the study area , D_0 represents a part of D_L after the location i is removed from D_L. So the conditional probability distribution is Gaussian :

$$\{Z_i \mid Z_j; j \in D_0\} \sim N(\mu \mid D_0, \sigma_i^2 \mid D_0)$$

Here: $\mu_i \mid D_0 = \gamma_{ij}^T \Gamma^{-1} Z_j^+$, $\sigma_i^2 \mid D_0 = -\gamma_{ij}^T \Gamma^{-1} \gamma_{ij}$, γ_{ij} is a $1 \times (N_0 + 1)$ vector containing $-\gamma_{|i-j|}$; $j \in D_0$ with the $(N_0 + 1)$th element of 1;

Γ is   a $(N_0 + 1) \times (N_0 + 1)$ matrix containing $-\gamma_{|j-k|}$; $j, k \in D_0$ with 1 in the last row and the last column except 0 in the location $(N_0 + 1, N_0 + 1)$;

Z_j^+ is a $1 \times (N_0 + 1)$ vector containing $Z_j; j \in D_0$ with 0 in the last location;

The realizations of permeability, porosity and shale percentage can be obtained by screening sequential algorithm, and will be used as input to simulate braided channels. The practical procedures are as follows.

CONDITIONAL SIMULATION OF BRAIDED CHANNELS

According to the reservoir characteristics of the study area, three previous geophysical parameters are used to determine the braided channel positions. The geophysical parameters always get smaller when depth increases, so these values should be calibrated with depth for the determination of channels.

Let Per, Por, Sh and H stand for permeability, porosity, shale percentage and layer depth, respectively. A discrimination value PP can be used to determine whether a 2D point belongs to a channel :

$$PP = \alpha_1 \times Per + \alpha_2 \times Por - \alpha_3 \times Sh \tag{1}$$

here: $\alpha_1, \alpha_2, \alpha_3$ are nonnative coefficients determined by geological knowledge, and depend on depth H. If $PP \geq Q$, the location belongs to a channel; If $PP < Q$, the location belongs to the sand between channels and shale; If Per=0.0, the location belongs to shale. Here, the value Q is a value determined by geological knowledge, and depends on depth H as well.

Based on formula (1), discrimination grid data $\{PP(i,j); j=1, ..., NY, i=1, ..., NX\}$ can be made. NX is the number of grid nodes in x-direction, NY is the number of grid nodes in y-direction. The PP values are used as input to simulate channel positions, and will be considered again when channel width is determined.

The procedures to simulate braided channel positions are discussed as follows. Firstly, each channel's center lines are simulated. Secondly, channel boundaries are obtained by widening the channel center lines. The procedures can ensure the simulated channels to pass the observed channels in the wells with probability 1.0. The connections and tributaries of channels follow geological knowledge.

Simulation of Braided Channel positions
The kernel technique is the simulation of braided channel positions. At the beginning, the starting point of each channel is searched in the study area, then the random walk model is used to find out the center line of a channel. The result is a series of grid nodes, among which the starting point is the first node.

The main factors considered here are follows: (1) well locations, (2) facies distributions presented by well data (channels, shale and sand between channels and shale), (3) PP values. Based on these, all the possible channels can be determined, the connections and tributaries of braided channels are considered as well.

Firstly, the information about facies in each well is assigned to a grid node which is the nearest one to the well location. An integer $KG(i, j)$ may have the following values:

$$KG(i, j) = \begin{cases} -1, & \text{if } (i, j) \text{ is the boundary location of grid system ;} \\ 0, & \text{if } (i, j) \text{ is not observation location ;} \\ 1, & \text{if } (i, j) \text{ is shale observation location ;} \\ 2, & \text{if } (i, j) \text{ is sand observation location between channels ;} \\ 3, & \text{if } (i, j) \text{ is channel observation location.} \end{cases}$$

Starting Positions of Channels
Let D_L be a grid system in the study area (Figure 1), ΔX and ΔY are the corresponding width of two narrow belts. Searching from (i,j); $i: NX \rightarrow NI, j: NI \rightarrow NY$ in one narrow belt along the west-east direction. If the first location (i_1, j_1) is found out where $KG(i_1, j_1)$ equals 3 and the next location (i_2, j_2) is obtained where $KG(i_2, j_2)$ equals 2 or 1, therefore i_1 can be considered as the x coordinate of the starting point of the first channel.

Similarly, searching from (i,j); $i: NJ \rightarrow NI$, $j:1 \rightarrow NY$ in another narrow belt along the north-south direction. If the first location (i_3, j_3) is found out where $KG(i_3, j_3)$ equals 3, therefore $j_3 - NJ$ can be marked as the y coordinate of the starting position of a channel.

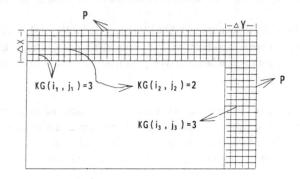

Figure 1 Finding the initial positions of channels

The starting points of all possible channels in the study area can be found out in sequence according to the previous procedures.

Random walk model in two dimensions will be used to determine a grid node whether emigrates from one direction among **a**, **b** and **c** (Figure 2).

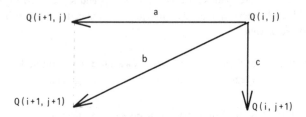

Figure 2 migration of grid node

Conditional Simulation of Flowing Locations and Bypassing Locations of the First Channel
Let the current location be Q(i,j), the determination of the next point relies on one of the direction **a, b** and **c**.

1. *Direction **a***
Emigrating from location Q(i,j) along the direction **a** is in order to find the nearest observed location (i_a, j), here i_a represents the corresponding nearest distance, Let Λ ="a location (i,j+1) is found, which satisfies $KG(i,j+1)=3$", If $P(Q(i,j) \rightarrow Q(i+1,j))$ represents emigrating probability, then

$$P(Q(i,j) \rightarrow Q(i+1,j)) = \begin{cases} 1, \text{ if } KG(i_a, j) = 3 \text{ or } \Lambda; \\ \alpha_1 \times (DA/DX) + \alpha_2 \times (PP(i+1,j)/MaxPP), \text{ if } KG(i_a, j) = 2, \\ \text{ and } KG(i+1,j) \neq 1 \text{ or } 2; \\ \alpha_3 \times (DA/DX) + \alpha_4 \times (PP(i+1,j)/MaxPP), \text{ if } KG(i_a, j) = 1, \\ \text{ and } KG(i+1,j) \neq 1 \text{ or } 2; \\ 0, \text{ other situations}; \end{cases}$$

here: $DA = |i_a - i| \times dX$; $DX = (NX-1) \times dX$; dX represents the distance between two adjacent grid nodes in x direction; MaxPP is the maximum of pp values in grid system; $\alpha_1, \alpha_2, \alpha_3, \alpha_4$ are nonegative parameters which are determined by geological knowledge, and $0 < \alpha_i < 1$, i=1,2,3,4.

If the next point has been found along the direction **a**, let KG(i+1,j)=3, otherwise considering the direction **b**.

2. Direction *b*

The direction **b** is toward the left-down direction, and is another way to find out the nearest observed location Q(i+1,j+1) in probability P(Q(i,j)->Q(i+1,j+1)) ;

$$P(Q(i,j) \rightarrow Q(i+1,j+1))$$

$$= \begin{cases} 1, \text{ if } KG(i_a, j_b) = 3; \\ \beta_1 \times (DB/\sqrt{(DX)^2 + (DY)^2}) + \beta_2 \times (PP(i+1,j+1)/MaxPP), \text{ if } KG(i_a, j_b) = 2, \\ \text{ and } KG(i+1,j+1) \neq 2; \\ \beta_3 \times (DB/\sqrt{(DX)^2 + (DY)^2}) + \beta_4 \times (PP(i+1,j+1)/MaxPP), \text{ if } KG(i_a, j_b) = 1, \\ \text{ and } KG(i+1,j+1) \neq 2; \\ 0, \text{ other situations}. \end{cases}$$

here: $DB = \sqrt{(i_a - i)^2 (dX)^2 + (j_b - j)^2 (dY)^2}$; $DX = (NX - 1) \times dX$; dX represents the distance between two adjacent grid nodes in X direction; $DY = (NY - 1) \times dY$; dY represents the distance between two adjacent grid nodes in Y direction; $\beta_1, \beta_2, \beta_3, \beta_4$ are nonegative parameters which are determined by geological knowledge, and $0 < \beta_i < 1$, i = 1, 2, 3, 4.

If the realization of a channel is migrating toward the direction **b**,that is $Q(i,j) \rightarrow Q(i+1,j+1)$, let KG(i+1,j+1)=3, otherwise considering the direction **c**.

3. Direction *c*

If the migrating direction is neither **a** nor **b**, then must be **c**, the pass is from Q(i, j) to Q(i, j+1), at the same time, let KG(i,j+1)=3.

The previous procedures are carried out repeatedly until KG(i,j)=-1, so the positions of the first channel are simulated.

Conditional Simulation of All the other Possible Channels
To simulate positions of other channels, the values of KG(i, j) are changed as follows:

$$KG(i, j) = \begin{cases} 4, & \text{if } (i, j) \text{ is the first channel location, but not an observation channel location;} \\ 5, & \text{if location } (i, j) \text{ is the first channel location and an observed channel location;} \\ KG(i, j), & \text{other situations.} \end{cases}$$

The connections and tributaries of channels will be considered in the following part.

1. *Direction* **a**
Let location Q(i,j) migrates toward direction **a** till the first channel location is searched which is (i_a, j), and Λ ="a location (i', j+1) is found from location (i, j+1) toward direction **a**,which satisfies KG(i',j+1)=3,and i" satisfies $i < i'' < i'$,KG(i", j+1)=1 or KG(i", j+1)=2". $P(Q(i,j) \rightarrow Q(i+1,j))$ represents the migration probability, then

$$P(Q(i, j) \rightarrow Q(i+1, j)) =$$
$$\begin{cases} 1, & \text{if } KG(i_a, j) = 3 \text{ or } \Lambda; \\ \gamma_1 \times (NX / (NX + DA)) + \gamma_2 \times (PP(i+1, j) / MaxPP), & \text{if } KG(i_a, j) = 5, \text{ or } 4, \\ & \text{and } KG(i+1, j) \neq 1, \text{ or } \neq 2; \\ \gamma_3 \times (DA / DX) + \gamma_4 \times (PP(i+1, j) / MaxPP), & \text{if } KG(i_a, j) = 2, \text{ and } KG(i+1, j) \neq 1, \text{ or } \neq 2; \\ \gamma_5 \times (DA / DX) + \gamma_6 \times (PP(i+1, j) / MaxPP), & \text{if } KG(i_a, j) = 1, \text{ and } KG(i+1, j) \neq 1, \text{ or } \neq 2; \\ 0, & \text{other situations.} \end{cases}$$

here: $DA = |i_a - i| \times dX$, DX and dX have the same meanings as the pervious ones; MaxPP is the maximum of pp values in grid system; $\gamma_1, \gamma_2, \gamma_3, \gamma_4, \gamma_5, \gamma_6$ are nonegative parameters which are determined by geological knowledge, and

$$0 < \gamma_i < 1, i = 1, 2, 3, 4, 5, 6;$$
$$\gamma_1 + \gamma_2 \leq 1, \gamma_3 + \gamma_4 \leq 1, \gamma_5 + \gamma_6 \leq 1;$$
$$\gamma_1 \geq \gamma_3 \geq \gamma_5, \ \gamma_2 \geq \gamma_4 \geq \gamma_6.$$

It is obvious that if the nearest searched location belongs to a channel, the bigger the migrating probability is, the smaller the nearest searched distance is. Therefore the channels will connect each other with more probability. Conditions $\gamma_1 \geq \gamma_3 \geq \gamma_5$ and $\gamma_2 \geq \gamma_4 \geq \gamma_6$ can show some characteristics that shale observation between channels will increase the chance of channel branching.

If the channel trace is from $Q(i,j)$ to $Q(i+1,j)$, when the location $Q(i+1,j)$ is not the channel location let $KG(i+1,j)$ equal 4, otherwise let $KG(i+1,j)$ equal 5. If the trace is not from $Q(i,j)$ to $Q(i+1,j)$, the direction **b** will be considered.

2. Direction **b**

Let the location $Q(i,j)$ migrate toward the left-down direction till the first location is found out which is marked as (i_a, j_b). If $KG(i+1,j)=1$ or $=2$, and $Q(i, j)$ does not migrates to $Q(i+1, j+1)$, so the migration probability from $Q(i,j)$ to $Q(i+1,j+1)$ is

$$P(Q(i, j) \rightarrow Q(i+1, j+1))$$

$$= \begin{cases} 1, \text{ if } KG(i_a, j_b) = 3; \\ \delta_1 \times (\sqrt{(DX)^2 + (DY)^2} / \sqrt{(DX)^2 + (DY)^2} + DB) + \delta_2 \times (PP(i+1, j+1) / MaxPP), \\ \quad \text{if } KG(i_a, j_b) = 4 \text{ or } = 5; \\ \delta_3 \times (DB / \sqrt{(DX)^2 + (DY)^2}) + \delta_4 \times (PP(i+1, j+1) / MaxPP), \text{ if } KG(i_a, j_b) = 2; \\ \delta_5 \times (DB / \sqrt{(DX)^2 + (DY)^2}) + \delta_6 \times (PP(i+1, j+1) / MaxPP), \text{ if } KG(i_a, j_b) = 1; \\ 0, \text{ other situations.} \end{cases}$$

here: $DB = \sqrt{(i_a - i)^2 dX^2 + (j_b - j)^2 dY^2}$; DX, DY, dX, and dY have the same meaning as the previous ones; $\delta_1, \delta_2, \delta_3, \delta_4, \delta_5, \delta_6$ are nonegative parameters which are determined by geological knowledge, and

$$0 < \delta_i < 1, \ i = 1, 2, 3, 4, 5, 6;$$
$$\delta_1 + \delta_2 \le 1, \ \delta_3 + \delta_4 \le 1, \ \delta_5 + \delta_6 \le 1;$$
$$\delta_1 \ge \delta_3 \ge \delta_5, \ \delta_2 \ge \delta_4 \ge \delta_6.$$

Because the trace of channels is from the north east to the south west, the parameters δ_i and γ_i must satisfy following relationships:

$$\delta_i \ge \gamma_i \quad i=1, 2, 3, 4, 5, 6.$$

If the channel trace is toward the direction **b**, that is $Q(i, j) \rightarrow Q(i+1, j+1)$, let

$$KG(i+1, j+1) = \begin{cases} 4, \text{ if the location } Q(i+1, j+1) \text{ is not the former simulated channel location;} \\ 5, \text{ if the location } Q(i+1, j+1) \text{ is the former simulated channel location;} \\ KG(i+1, j+1), \text{ other situations.} \end{cases}$$

$$(2)$$

otherwise, the direction **c** will be considered.

3. Direction **c**

If the trace is neither along **a** nor **b**, it must be **c**, that is $Q(i, j) \rightarrow Q(i, j+1)$. At the same time, the value of KG(i,j+1) must be changed as (2).

In order to find out more branch channels the previous procedures are repeated until KG(i,j)=-1, the loop number of the procedure depends on practical geological characteristics of the study area. Generally, the more the number of searched branch channels is , the more frequent the connections and tributaries of channels are . In the study, only one branch channel is searched for each channel.

Simulation of Other Channels and Their Branch Channels
Similar method is used for finding other channels and their branch channels.

DETERMINATION OF CHANNEL BOUNDARIES

The width of each channel depends on PP values. The bigger the PP values are, the wider the channel is.

Pretreatment
If the trace of a channel is $M \rightarrow N \rightarrow L$, location N should be removed. It is obvious that the treatment can simplify the process of widening channel, but does not change the trace of each channel (Figure 3).

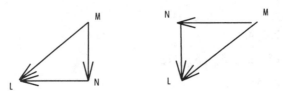

Figure 3 pretretment of widening channels

Determination of Channel boundaries
The formula of channel width is as follow:

$$\text{Width} = \Delta_1 + PP(i, j) \times \Delta_2 \Big/ \text{MaxPP}$$

here: Δ_1 is the minimum of channel width in the study area; Δ_2 is the maximum of channel width; PP(i,j) is the PP value of the nearest grid node which is adjacent to channel location. (Figure 4)

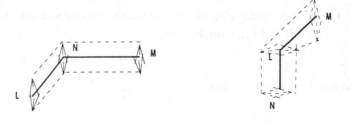

Figure 4 widening channels

A CASE STUDY

The study area Sheng-84 is located in LiaoHe Oil Field in China.

In this area, 100 well data are used, including petrophysical parameter information such as permeability, porosity and shale percentage, and facies information such as channels, shale, and sand between channels and shale. The realization of three geophysical parameters are obtained by fractal Brownian motion model.

Based on these simulations of geophysical parameters, facies realization can be produced by Random Walk Model (Figure 5).

The parameters of the model are chosen as follows:

NX=NY=65; DX=50m,DY=30m;
$\alpha_1 = 0.2,\ \alpha_2 = 0.3,\ \alpha_3 = 0.1,\ \alpha_4 = 0.2;\qquad \beta_1 = 0.3,\ \beta_2 = 0.4,\ \beta_3 = 0.2,\ \beta_4 = 0.3;$
$\gamma_1 = 0.2,\ \gamma_2 = 0.3,\ \gamma_3 = 0.1,\ \gamma_4 = 0.2,\ \gamma_5 = 0.1,\ \gamma_6 = 0.2;$
$\delta_1 = 0.3,\ \delta_2 = 0.5,\ \delta_3 = 0.3,\ \delta_4 = 0.4,\ \delta_5 = 0.2,\ \delta_6 = 0.3; \Delta_1 = 70m,\ \Delta_2 = 50m.$

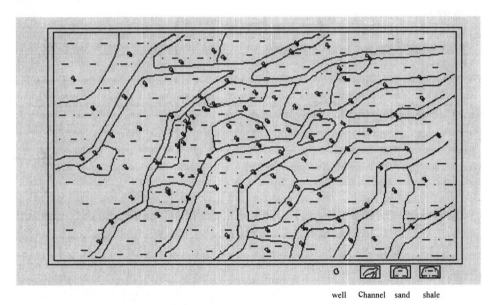

well   Channel   sand   shale

Figure 5 one simulation of braided channels

CLOSING REMARKS

In deltaic sedimentary environment, the reservoir controlled by braided channels has great heterogeneity, because of narrow width of channels and frequent connections and tributaries of channels. It is very important to characterize the distributions of braided channels in two dimensions.

In this paper, random walk, as a stochastic modeling method in two dimensions, is constructed to characterize the distributions of braided channels. The realizations produce some important characteristics of braided channels: frequent connections and tributaries of channels. At the same time, the width of channel is considered in the model, and the continuity and smoothness of channels are reserved.

ACKNOWLEDGMENTS

Special thanks to Rongzhi Zheng and Huanpeng Li of the Geological Science Research Institute of LiaoHe Petroleum Cooperation for their technical support.

Special thanks to Prof. F.P.Agterberg for his beneficial suggestion of the revision of the paper.

REFERENCES

1. Haldorsen,H.H. A new approach to shale management in field scale simulation models, *SPE(10976)*, 447 ~ 457 (1984).
2. Matheron,G.,Beucher,H.,de Fouqqet, C., Galli,A., Guerillout, D.and Ravenne,C. Conditional simulation of the geometry of fluriodeltaic reservoirs, *SPE 62nd Annual Conference Dallas, Texas*, 591 ~ 599 (1987).
3. Olivier Dubrule. A review of stochastic models for petroleum reservoirs, *GEOSTATISTICS.* **2**, 493 ~ 506 (1989).

Table 2. Typical examples of high speed for real-time cartography

name or content	usage	specification	time (days)
geological, engineering, and seismic zoning maps for dame site of Three Gouges in Yangtse River(1992)	for assessing dame site of Three Gouges prepared by committee of water conservancy in Yangtse River, used by the State Council and more than ten ministries and committees	A0 format 3 maps	10
transportation, commerce and trade maps for Shen Zhen city (1993)		when the map was ready to printing, customer demands to change all numbers of buses of the city within 10 days	> 1
the national maps of agriculture, water conservancy and resources in 1:4000000 (1994)	for central government	double A0 9 maps	12
travel resource maps of Hu Bei province (1994)	for economic and trade exhibition hold in Germany and France, by Hu Bei province government	double A0, 5 maps in both Chinese and English	5
sugar development planning map of Guang Xi province (1995)	present to The State Council made by planning committee of Guang Xi province	format: 2.5 x 1.7 (m)	7

example, in 1:50000 regional geological mapping program which is the most fundamental geological mission and there are several hundreds maps need to be published, through studying and testing, new production procedure for digital technology has been proposed and several graphbase of legend, patterns, line types, symbols and color code are being developed based National Geological Mapping standard. At least 100 maps have been digitized.

By the end of July of 1996, more than 50 companies are formed by using the system to digitizing various maps. It can be said that the application of the system drive a new enterprise to take shape.

Development and Application of Geographical Information System - MAPGIS[12] has a Bright Future

As everyone knows, GIS is a powerful tool for geoscience experts. With GIS, the experts can analyze and integrate multiple-source information from geology, geophysics, geochemistry, remote sensing and etc. to research   various geoscientific problems.

Experiences in both external and domestic application of GIS have made us understand very well that the special significance of GIS in helping us to get rid of the traditional work way, and application demand in MGMR. The project to develop GIS tool software was lunched in 1992, which is named MAPGIS. The project is great successful and MAPGIS on PC has released in 1995 with the following features.

- It is developed with c++ on Windows, with user friendly and easy-to-use interface;
- Map can be digitized by both digitizer and scanner, and provide perfect error correcting function;
- Dynamic definition and editing of topographic relations and attributes;
- Mapbase management system can manage more than thousands of maps and is powerful for retrieval across these maps and clipping and pasting up;
- With both vector and raster data structures and data with different structure can be effectively converted;
- Powerful capability of spatial data analyzing including 2.5D data analyzing and displaying;
- Function library is power tool for application development;
- Interface to ARC/INFO, INTERGRAPH, DBASE, and FOXBASE;
- Drivers for all kind output devices, and Post Script interface for output of plate film.

Two application projects (mineral resources evaluation system and tubenet management system) have finished, and several other application projects, such as emergency reply system, city planning and land use management system are developing. MAPGIS has bright application future!

GEOSCIENCE INFORMATION SYSTEMS WILL HAVE A BIG PROGRESS DURING THE NEXT FIVE YEARS

Requirement
- Informatization program of geology and mineral resources field has set up by the headquarters of the ministry and will be a priority project to develop. The main goal of the program is the following: Realizing informatization of main procedures in geological and mineral resources survey and administration management and geoscience achieve management. To achieve the goal a number of geoscience information system need to be built up and corresponding computer aided tools need to be developed.
- By the end of 1995, 3250 sheets in 1:50000 scale have covered 1400000 KM^2, making up 14.6% of the whole country. During nest five years, there are more than 800 sheets need to be mapped.
- In order to reasonably develop and utilize groundwater resource and reduce geological hazard, two programs are going to start. One is 1:100000 groundwater development and utilization surveying in 600 counties, the other one is 1:500000 geological environment surveying in 17 provinces.
- The national geological archives saves 85000 copies various kinds of geoscience reports, which are full of the all 6 floors of the building. Optical disk technology is considered a good way to solve the problem.

Important Development Projects
Developing regional integrated geoscience information system focusing on spatial information like digital geoscience maps in different scales and integrated databases.
- Computer aided 1:50000 geological mapping system;
  Goal of the project is to develop an system and work flow for the whole procedure from data acquisition in the field, data management, data integration, color map production based on 3S (GIS, GPS and RS) technology.
- Computer aided water resource and geological environment surveying system;
- Decision support system of mineral resources evaluation ;
- Management system for exploration and development of mineral resources;
- Geoscience achieve storage and retrieval system with optical disk.

Key Points for Realizing the Goal
- Technology: GIS , GPS, RS, field data acquisition; facilities and multi-media;
- Standardization;
- Application development;
- Training ;
- Coordination and management.

REFERENCES

1. Anonymous. Development of National Geochemical Reconnaissance Database (NGCHRD). *internal report* , Geological Survey of MGMR (1996).
2. Anonymous. Development of National Gravity and Elevation Database (NGED), in*ternal report* , Geological Survey of MGMR (1996).
3. Anonymous. Development of Aeromagnetic Database (ARMD). in*ternal report.* Geological Survey of MGMR (1996).
4. Chuanlin Chen and et al.. Petroleum Geology and Drill Hole Database (PGDH). *internal report.* bureau of Petroleum Geology, MGMR (1993)
5. Chaoling Li, Yi Jiang and et al.. China Stratigraphical Information System (STRL). *research report.* Geological Survey of MGMR (1995).
6. Zuoqin Jiang and et al.. MAPCAD application investigation report. internal *report.* Dep. of Science and Technology, MGMR (1996).
7. Deyao Ma, Zhongyu Wu and et al.. Geoscience Database Directory (GDBD). *research report.* Dep. of Science and Technology, MGMR (1995).
8. Yingji Pan, Guizi Lin and et al.. Development of Mineral Exploration Information System (MAGAD). research *report.* Dep. of Science and Technology, MGMR (1993).
9. Kuanlian Wu, Banggong Cao and et al.. Fundamental Geoscience Graphic Information System (FGGI). *research report.* Dep. of Science and Technology, MGMR (1995).
10. Kuanlian Wu, Jinsheng Yu and et al.. Fundamental Geoscience Software Library (FGSL). *research report.* Dep. of Science and Technology, MGMR (1993).
11. Xincai Wu and et al.. Development of Computer Aided Editing and Publishing System for Color Geoscience Maps - MAPCAD. *research report. research report.* Dep. of Science and Technology, MGMR (1993)
12. Xincai Wu and et al.. Development of General Geographic Information System-MAPGIS. *research report.* Dep. of Science and Technology, MGMR (1995)
13. Rongfu Xia, Bixing Jiang . Development of the Second Version of National Mineral Reserve Data Base (NMRD). *situation and challenge.* 451-476 (1994*).*
14. Binxung Yang and et al.. The National Geological Library Management System(NGLMS). *research report.* Dep. of Science and Technology, MGMR (1995).
15. Qingdi Yang and et al.. Development of Computerized Mineral Exploration and Evaluation System (KPX). *research report.* Information Institute of Geology and Mineral Resources of CHINA (1994).
16. Jingman Zhao, Deyao Ma and et al. .Speeding Development of Geological and Mineral Information Systems to Promote Modernization of Geology and Mineral Resources . *situation and challenge.* 477-497 (1994*).*

Proc. 30th Int'l Geol. Congr., Vol.25, pp. 133-144
Zhao Peng-Da et al (Ed.)
© VSP 1997

3D Visualization of Rock Textures

TAKAO ANDO, SOICHI OMORI, YOSHIHIDE OGASAWARA

Institute of Earth Science, Waseda University, Shinjuku-ku, Tokyo 169-50, Japan

J.B. NOBLETT

Department of Geology, Colorado College, Colorado Springs, CO 80903, USA

Abstract

Computer tomography technique by serial-grinding has been developed for understanding 3D rock textures. Rock samples that were cut into several cm cube were sequentially ground and were taken in color pictures at every 0.5 mm thickness by hand-operation. Based on a set of 2D section images derived from serial section pictures, 3D rock textures were synthesized by a visualization software "AVS" in two ways; cross-section method and volume-rendering method. This technique has been applied to two kinds of rock sample; metamorphosed composite intrusive rock showing complicated fluidal texture of mafic and felsic parts from the Hida metamorphic belt, Japan, and garnet porphyroblast-rich eclogite from Franciscan terrane, California. For the former sample, the resultant 3D textures show commingling relation between mafic and felsic parts and strongly suggest the mingling of two magmas before metamorphism. For the latter sample, the 3D images clearly show the size, amount and distribution of garnet porphyroblasts, and such data may contribute to discuss the nucleation of garnet porphyroblast in eclogite. The serial-grinding CT with the 3D visualization software is available for understanding real rock textures ranging from several mm to cm scale by use of sample block of several cm size cube.

Keywords: 3D visualization, Rock texture, magma mixing, eclogite

INTRODUCTION

Recent progress in software and hardware for 3D image processing has greatly improved the methods to understand many kinds of phenomena and objects in several special fields. Computer Tomography (CT) is one of the best successful method and has been already applied to medical sciences as X-ray CT.

In petrology, rock textures have usually been recognized as two-dimensional section images by the naked eye and under the microscope, and then petrologists have to imagine or to reconstruct real 3D textures on the basis of 2D information. 3D images of rock textures contain much information on petrogenesis; however, we cannot directly observe inner texture of real rock samples. It has been known that too much reliance on 2D textural information leads to misinterpretation [1,

2]. To clear such difficulties in rock textures, the 3D visualization technique with graphic computer is much expected; however, its application seems to be hard. The pioneer works of 3D X-ray CT for rock textures have been given by Prof. W.D. Carlson's group at the University of Texas [e.g., 3, 4]. They have already made almost fully automatic X-ray CT system for rock textures. Using the 3D visualization software AVS, we also have tried to establish the 3D observation technique by serial-grinding and serial-section pictures [5, 6, 7, 8]. This technique has almost been completed and may be available for several rocks.

The purpose of this paper is to describe "serial-grinding CT" technique to get 3D image of rock textures as two examples of composite intrusive rock and garnet porphyroblast-rich eclogite, and to show the resultant synthesized 3D textures of those rocks.

SAMPLE DESCRIPTION

Metamorphosed composite intrusive rock
Metamorphosed composite intrusive rocks that were utilized for 3D observation were collected at the Higashi-Urushiyama outcrop, Gifu Prefecture, Hida metamorphic belt, central Japan. This rock consists of two parts; melanocratic part of amphibolite, and leucocratic part of metatonalite. This composite rock, which has been subjected to later stage Hida metamorphism, has intruded into the host rock of biotite-hornblende gneiss that has been subjected to earlier stage metamorphism. As shown in Fig. 1a, this rock shows complex mingling and fluidal textures. Such textures seems to suggest the mingling of mafic and felsic magmas. On the basis of Rb-Sr isochron age dating, however, [9] reported the age difference of ca. 82 Ma between leucocratic and melanocratic parts, and he regarded the leucocratic part as a younger intrusive into the older mafic intrusive rock, although his age data have large uncertainties. If this rock has been formed by magma mingling, the age difference should be regarded as error. To observe the real 3D textures may contribute to clarify the genesis of this composite texture. The specimen for 3D observation was cut into $6 \times 10 \times 12$ cm parallelopipedon (Fig. 1a).

Eclogite
The eclogite was collected at Jenner in Franciscan Terrane, northern California, USA. This rock occurs as tectonic blocks in the melange zone [10, 11]. The principal constituent minerals are garnet, omphacite, phengite, glaucophane, chlorite. A large amount of garnet porphyroblasts occurs in this rock. Garnet crystals that usually have a crystal form of euhedral dodecahedron are 2~9 mm in diameter, and show reddish brown color to the naked eye. Under microscope, garnet crystals are often replaced by chlorite along the rim. The matrix of this eclogite mainly consists of omphacite and glaucophane that is a product of retrograde stage. Omphacite-rich part shows dark green color and glaucophane-rich part shows bluish dark gray. The main purpose of 3D observation of this sample is to understand the size, amount and distribution of garnet porphyroblasts in eclogite. The specimen for 3D observation was cut into $5 \times 10 \times 15$ cm parellelopipedon (Fig. 1b).

Figure 1. Photographs of real rock samples used in this study. a: metamorphosed composite intrusive rock from the Hida metamorphic belt, Japan. b: garnet porphyroblast-rich eclogite from Franciscan Terrane, California.

METHOD OF 3D IMAGE SYNTHESIS

Sample preparation and image acquisition
The working flow from photograph acquisition to the preprocessing of serial images is summarized in Fig. 2. Samples for this observation were cut into a rectangular parallelepiped in the size about several cm cube. Photographs of all six surfaces of the parallelopipedon were taken before serial grinding in order to compare those with the synthesized surfaces. After the surface for grinding was specified, it was ground with #100, 400, 800 abrasives to be shaved at a 0.5 mm interval, and was taken as a color picture of negative film. Such procedures were repeated two hundred times. The error of one hundred times shaving was less than 1 mm.

Hardware and software
Two kinds of UNIX workstations, Titan3000 manufactured by Stardent Co. Ltd. and Magnum4000 by MIP Co. Ltd. were used in this study. The 3D visualization software, AVS (Application Visualization System) distributed by Kubota Graphics Technologies Inc., was used for the synthesis of 3D images. AVS is one of the famous 3D visualization software available for many kinds of computers. The characteristic features of AVS are the interactive user interface and the four subsystems; image viewer, graph viewer, geometry viewer and network editor. The network editor gives us a visual programming environment to make some application programs only by connecting executable modules with mouse operation.

By use of the network editor, two kinds of applications were made for 3D observations; one is cross-section method and other is volume-rendering method.

Preprocessing of serial 2D images
Color photographs of serial sections were used for 3D observation of rock textures. Each photograph was read with an image scanner in the resolution of 75 dpi and was stored as a file of X-Window Dump File format (xwd). Each xwd file was converted to 2D field data format in AVS. Then, a set of serial section images was converted to 3D voxel data for AVS by our original program "2D_to_3D" that is written in C.

Observation by cross-section method
This method generates an arbitrary cross-section of a synthesized 3D object at any directions and creates a continuous section like an animation. Figure 3a shows the AVS module network to observe 3D textures by cross-section method. The function of each module is as follows: "read field" reads AVS field data, "crop" changes the size of field data, "downsize" changes the size of 3D voxel data, "generate colormap" produces an AVS colormap data structure which is used by "colorizer" transforming input voxel data to color values, "brick" creates a picture of volume data, "animated float" antomatically modifies the parameters for animation.

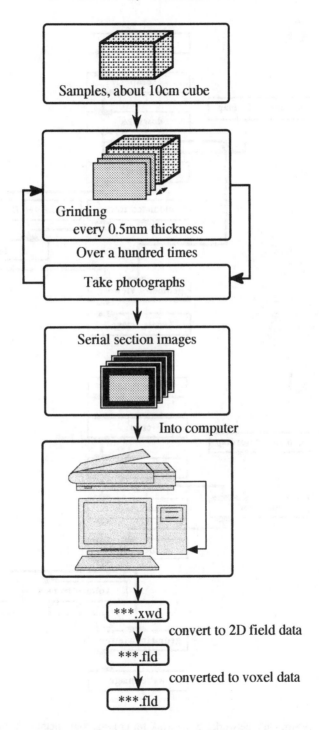

Figure 2. Working flow from photograph acquisition to the preprocessing of serial section images.

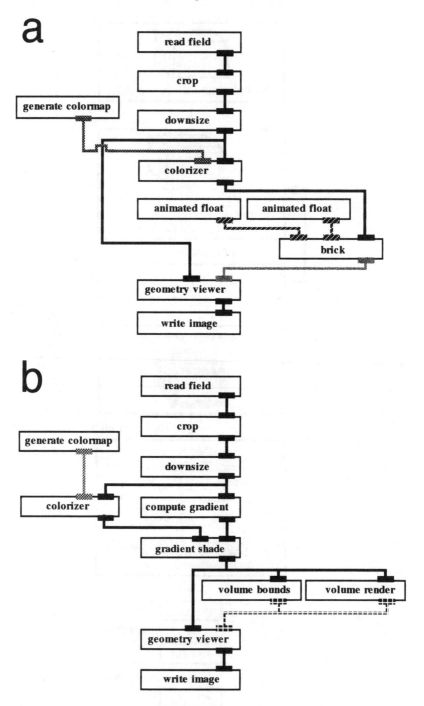

Figure 3. AVS module networks. a: network for cross-section method, b: network for volume-rendering method.

Observation by volume-rendering method
This method generates transparent 3D images. Figure 3b shows the AVS module network to observe 3D textures by volume-rendering method. Until "downsize" module, the process follows the cross-section method. The functions of the other modules are as in the following: "compute gradient" and the "gradient shade" compute the gradient vector at each point in a 3D field data set, "volume bounds" generates lines that indicate bounding box of a 3D field data set.

RESULTS AND DISCUSSIONS

Metamorphosed composite intrusive rock
The purpose of this sample was to observe the real 3D texture, particularly to detect the commingling texture between leucocratic part and melanocratic part. The synthesized 3D textures both by cross-section and volume-rendering methods were successfully illustrated in the synthesized images (Fig. 4) which are originally represented as black and white gray scale images. The interval of shaving, 0.5 mm thickness, is sufficient for synthesizing 3D images of several cm size rock specimens. Figures 4a to 4c show the synthesized 3D texture by cross-section method. All six surfaces of the real rock sample (Fig. 1a) are clearly reillustrated in the synthesized 3D image (Fig. 4a). Examples of cross sections at arbitrary directions are shown in Figs. 4b and 4c. By observing continuous cross sections at an appropriate direction on monitor screen like an animation, we confirmed the commingling relation between melanocratic and leucocratic parts: some isolated leucocratic parts are included in melanocratic part, and some melanocratic parts in leucocratic part (Fig. 4c, arrow A).

Figures 4d to 4f illustrate the 3D form of leucocratic part generated by volume-rendering method. Melanocratic part is set as completely transparent. The brightness of gray scale image has been employed to discriminate the leucocratic part from melanocratic part. For discriminating objects from other portions in a 3D image, this rock is simple case because this consists of two parts and it is sufficient to divide into two parts by setting an appropriate threshold of the brightness.

The shapes of leucocratic part show "pipe-like" and "branch-like" forms in melanocratic part. No leucocratic part shows "plate" form. The tiny leucocratic part is branching from the relatively large branches, and some show "droplet-like" form as isolated parts in melanocratic part (Fig. 4f, arrow B). In order to explain such texture, we have to introduce the magma-mingling between felsic and mafic magmas. If the origin of leucocratic part is later stage intrusion into melanocratic part that was already consolidated, its form should be a "plate". The forms of leucocratic part suggest that the melanocratic part had been a liquid when felsic intruded into it. These 3D shapes of the leucocratic part obtained in our 3D method may strongly suggest that the meta-composite rock at Higashi-Urushiyama outcrop is formed by mingling of mafic and felsic magmas as already reported in Colorado, USA [e.g., 12].

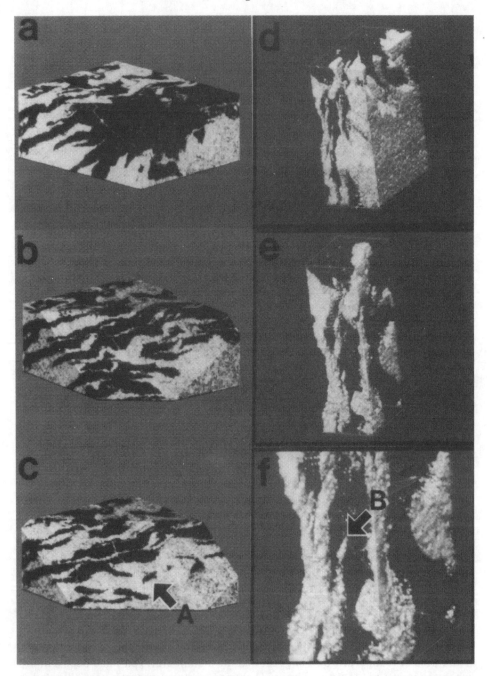

Figure 4. Synthesized 3D images of the metamorphosed intrusive rock from the Hida metamorphic belt, Japan. a, b, c: 3D images by cross-section method. Dark parts are mafic part, and light parts are felsic part. d, e, f: 3D images of felsic part by volume-rendering method. Mafic part is set as transparent.

Franciscan eclogite

The purpose of this sample was to understand the size, amount and distribution of garnet porphyroblasts in eclogite. Synthesized 3D images both by cross-section method and volume-rendering method are shown in Fig. 5. An arbitrary section of the cube is illustrated in Fig. 5b. Being compared with the photograph of the eclogite sample (Fig. 1b), all six surfaces of the synthesized cube are well reproduced (Fig. 5a).

In this paper, the printing of 3D texture of eclogite is represented in black and white although the original synthesized 3D images have been drawn in full-color. Light parts in each photograph in Fig. 5 indicate garnet porphyroblasts. Darker parts are the eclogite matrix mainly consisting of omphacite and glaucophane. The garnet porphyroblasts which are relatively coarse-grained crystals (>6 mm) were well discriminated from the eclogite matrix by the difference of color tones, but small-grained crystals (<2mm) were slightly difficult to be discriminated (Fig. 5b, arrow). Also, it is very difficult to discriminate other minerals in the matrix. Major constituents in matrix are omphacite and glaucophane. Those colors are dark green and dark blue, respectively; so, it is very difficult to discriminate each other by color picture of polished surface. Some special preprocessing to specimen surface or to serial-section images should be required to discriminate those minerals.

The grain size variation is easily recognized in 3D images, but the crystal shape of garnet porphyroblasts is not illustrated in those photographs due to the low resolution derived from 0.5 mm shaving interval. The distribution of garnet porphyroblasts has been well observed in the synthesized 3D images, particularly in those by volume-rendering which were obtained by setting the matrix of eclogite as completely transparent under AVS operation (Figs. 5c, d). Relatively small-grained garnets are concentrated in the lower layer of the specimen (Fig. 5d, arrow); but large-grained garnets are mainly found in upper part and those distribution is a little sparse. Together with the bulk chemical composition data of each layer, 3D distribution data on garnet will contribute to consider the nucleation of garnet porphyroblasts in eclogite though this paper will not mention about this any more.

For another application of 3D texture, we tried to measure a modal composition of garnet in eclogite. The 3D modal value is 24.5 vol. %. This value was calculated from a set of 2D serial-section images. Ordinary point-counting method under the microscope gave 22.5 vol. % as garnet modal composition. This was obtained from three thin sections of other piece of the same sample that looks similar. The value of 3D method may be more reliable than that by point-counting method. Cutting specified portion of a 3D image, we can calculate the garnet modal composition of corresponding part of the sample.

CONCLUSIONS

The technique of 3D observation of rock textures by serial-grinding method with 3D visualization software AVS has been proposed. This method is available for

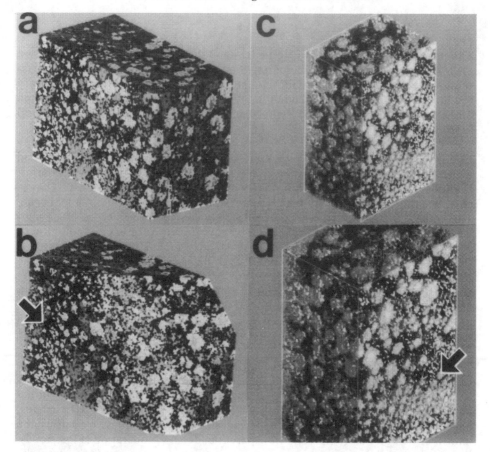

Figure 5. Synthesized 3D images of the garnet porphyroblast-rich eclogite from Franciscan Terrane, California. a, b: 3D images by cross-section method. Light parts are garnet porphyroblasts. Dark parts part the matrix mainly consisting of omphacite and glaucophane. c, d: 3D images garnet porphyroblasts by volume-rendering method. The matrix is set as transparent.

understanding real rock textures ranging from several mm to cm scale by use of sample block of several cm cube.

For the sample of metamorphosed composite intrusive rock, mafic and felsic parts were successfully discriminated each other. The observation of animated 3D images by cross-section method confirmed the commingling relation between mafic and felsic parts. The 3D images of felsic part obtained by volume-rendering method indicate the fluidal texture such as pipe-, branch- and droplet-like forms that suggest the felsic magma flow into unconsolidated mafic magma.

For garnet porphyroblast-rich eclogite, garnet crystals are well distinguished from the matrix by the difference of color tone, and the size and distribution of garnet porphyroblasts are easy to recognize in 3D images, particularly in those by volume-rendering. The 3D images are also available for the modal analysis of garnet in eclogite.

Our "serial-grinding CT" method for rock textures using 3D visualization software AVS is effective to the rock sample having several mm to cm scale texture in the size of several cm cube though this has some faults as follows: 1) this method requires high-level technique for rock-grinding, 2) after the acquisition of a set of serial-section images real rock samples have gone. The precise grinding technique in sub-mm scale developed in this study will be applicable for the synthesis of 3D chemical compositional images of garnet porphyroblast with electron microprobe in order to detect the real 3D compositional zoning.

Acknowledgments

We express our sincere thanks to Mr. Tatuya Nikkuni at KGT. Co. Ltd. for his kind advice on the usage of AVS.

REFERENCES

1. D. Shelly. *Igneous and metamorphic rocks under the microscope,* Chapman and Hall, 445 (1993).
2. M.J. Hibbard. *Petrography to petrogenesi,* Prentice Hall, 587 (1995).
3. W.D. Carlson and C. Denison. Mechanism of porphyroblast crystallization: results from high-resolution computed X-ray tomography, *Science* **257**, 1236-1239 (1992).
4. C. Denison, R.A. Ketcham and W.D. Carlson. Using high-resolution computed X-ray tomography for three-dimensional quantitative textural analysis (abs.), *Geol. Sci. Amer. Abs. with Progs.* **28**, A-54 (1996).
5. T. Ando, S. Omori, Y. Ogasawara and J.B. Noblett. Applications of 3D visualization technique to petrology-An example of magma mingled rock (abs.), *AGU Eos Abs. with Progs.* **75**, F696 (1994).
6. T. Ando, S. Omori, Y. Ogasawara. Synthesis of 3D textures of garnet porphyroblasts in Franciscan eclogite, *AGU Eos Abs. with Progs.* **76**, F669 (1995a).
7. T. Ando, S. Omori, Y. Ogasawara and J.B. Noblett. 3D observation of rock textures with AVS-An example of magma-mingled rock (in Japanese with English abstract), *Bull. Centre for Informatics, Waseda University* **20**, 12-23 (1995b).
8. T. Ando, S. Omori, Y. Ogasawara and J.B. Noblett. 3D observation of rock textures (abs.), *30th Int'l. Geol. Congr. Abs. with Progs.* **3**, 466 (1995).
9. Y. Arakawa. Rb-Sr ages of the gneiss and metamorphosed intrusive rocks of the Hida metamorphic belt in the Urushiyama area, Gifu prefecture, central Japan, *Japan Jour. Assoc. Min. Petrol. Econ. Geol.* **79**, 431-442 (1984).
10. R. Coleman and M.A. Lanphere. Distribution and age of high-grade blueschists associated eclogites, and amphibolites from Oregon and California, *Geol. Soc. Amer. Bull.* **164**, 77-91 (1971).
11. M.Cloos. Blueschists in the Franciscan complex of California: petrogenetic constraints on uplift mechanism. *Blueschists and Eclogites, Geol. Sci. Amer. Memoir.* **164**, 77-93 (1986).
12. J.B. Noblett and M.W. Stab. Mid-Proterozoic lamprophyre commingled with late-stage dikes of the anorogen San Isabel batholith, Wet Mountains, Colorado, *Geology* **18**, 120-123 (1990).

- Powerful edition functions. Representing complex spatial relations, for example, to multiple inlaid polygons, easy to deal with their spatial positions and colors correctly ;
- Capability to define line types, patterns and icons with parameters arbitrarily, rich data bases of line types, patterns, legends, Symbols and color codes based on national geological mapping standards;
- Data format compatible with ARC/INFO, INTERGRAPH and AUTOCAD;
- POSTSCRIPT interface for map publishing;
- Run on 486 PC with 8M memory on both DOS and WINDOWS.

Application of the System has Achieve a Great Success
A lot of digital cartography practice has approved the great advantages of new technology[6]:
- reducing processes of color map publishing from 12 steps to 5;
- Increasing productivity: for new map, It is 3-5 times as fast as the traditional method, for revising map, 10-20 times;
- Reducing cost : average 1/3
- Data sharing: digital data can be used repeatedly.

Now, the system is used wider and wider. About 700 sets of the software have been installed all over the country accept Taiwan and Tibet since the system was released in 1993. Some copies have been output to North Korea, and many other countries are interesting; There are more than 20 applying fields involved, which are not just geoscience fields, such as geology, petroleum, coal, mining, hydrogeology, land use and management, but also other fields: transportation, railway, city planing and management, environment, travel, and so on.

More than ten thousands of various color geoscience maps are produced by the system. These maps include the first set of atlases with digital cartographic technology such as The National Minerals Atlas, Land Use Atlas of GANJINGZI district of DALIAN, Soil Atlas of HEBEI province, The National Highway Atlas, Regional Geological and Geophysical Atlas of Sea Area of CHINA and etc. A series important maps from some national scientific research program. Among them the biggest one is The Aria and Europe Geological Map published during the 30th International Geological Congress with 9 sheets of extra A0 formats, the complex one has several thousands graph elements(polygons).

Application of the system has coursed a great technology progress in color map editing and publishing. It is shown in the following respects.

It can finish the emergency mapping task at a very high speed which is impossible and can't image in traditional manual method. Table 2[6] illustrates this by some typical practice examples.

It is making the traditional cartography change its working way in geosciences. Now, the digital cartographic technology has been adopted in geoscientific map production. For

Table 1. The major geoscience databases

name	time	content	system platform	volume
NMRD	1989	deposit information :geography, geology, minerals, reserves, grade, mining condition and etc.	SUN workstation, UNIX, ORACLE	18000 deposits, 65000 records and 100M
NGCHRD	1979	sampling information and analyzing data of 39 elements per sample	PC , DOS and WINDOWS	1 sample / 4KM^2 , covering 5.17 million KM^2
NGED	1985	5' x 5' and 1KM x 1KM elevation data, gravity data	PC , DOS and WINDOWS	elev. covering the whole country, grv. 60%
STRL	1995	name, classification, strata section description, stratigraphic column correlation, reference	PC 486, WINDOWS	about 5000 stratigraphic units, 300000, 210M
FGGI	1995	information about topography, hydrographic net, transportation, administration boundary and etc.	SUN workstation , UNIX, home made software	514 pieces of maps, 20M
GFSD	1993	basic mathematics, geophysics and geochemistry , geomathematics, image processing, hydrogeology, fundamental software	PC 486, DOS, MS - C, MS - FORTRAN	512 program modules, 120000 lines
GDBD	1995	name, content, coverage, volume of data, software and hardware, connect person	PC 486, WINDOWS, FOXBASE	1000 records

Proc. 30th Int'l Geol. Congr., Vol.25, pp. 125-132
Zhao Peng-Da et al (Ed.)
© VSP 1997

The State and Development of Geoscience Information Systems in the Ministry of Geology and Mineral Resources of CHINA(MGMR)

JIANG ZUOQIN

Technology Division, Dep. of Science and Technology, MGMR ,CHINA

Abstract

Geoscience information system is foundation of geoscience modernization and a very important part of national informatization infrastructure. Development of geoscience information system within MGMR started in the end of 1970's. During the longer than 20 years, a number of geoscientific databases in various fields are built up; A Computerized Mineral Exploration and Evaluation System(KPX) and Computer Aided Editing and Publishing System for Color Geoscience Maps - MAPCAD are developed in house, and are distributed and used all over the country, not only in MGMR, but in many other ministries. MAPCAD makes the whole procedure of color map publishing, from digitizing, editing, processing, proofing , color separating until output of plate film , be computerized on PC. It has achieved remarkable success in changing traditional working way; MAPGIS, as a general and powerful GIS tool software, also are developed in house and has been released. Several pilot research projects of GIS application with MAPGIS, including mineral resources evaluation, water resources evaluation, geological Hazard, geological cartography, land use and so on, are underway.

The paper consists of two parts. Part one is about the state and features of geoscience information systems, part two is development of the systems in the future.

Keyward: geoscience information systems, database, computer aided mapping, digital cartography, geographic information system (GIS)

THE STATE AND FEATURES OF GEOSCIENCE INFORMATION SYSTEM IN MGMR

A Number of Geoscientific Databases are Built Up
Through the effort in last more than 20 years, we have built about 100 databases[16] in geological literature, geochemistry, geophysics, geology, mineral resources and exploration, water resources and geological environment, management of mineral exploration and development, fundamental geological research and etc. The following are some major databases from coverage area and data volume: The National Mineral Reserve Data Base (NMRD)[13]; National Geochemical   Reconnaissance Database (NGCHRD)[1], National Gravity and Elevation Database (NGED)[2], Aeromagnetic Database (ARMD)[3], Stratigraphic Lexicon (STRL)[5], Mineral Exploration Information

System (MAGAD)[8], Fundamental Geoscience Graphic Information System (FGGI)[9], Petroleum Geology and Drill Hole Database (PGDH)[4], The National Geological Library Management System(NGLMS)[14], Geoscience Database Directory (GDBD)[7], Fundamental Geoscience Software Library (FGSL)[10].

Some main information about these databases illustrates in table 1.

Computerized Mineral Exploration and Evaluation System(KPX) [15]Has Become Effect
    Development of the system which is founded by both UN and the ministry was finished in 1993. The function of the system includes : data acquisition in the field (drill hole logging, drilling , engineering, and so on), data management, ore body correlation, reserve evaluation and exploration map drawing. Features of the system are as the following.
- the standardization schema is developed, which includes data dictionary for mineral exploration and data acquisition format;
- pattern graph can be created automatically;
- drawing complex columns of drill holes;
- support both traditional and geostatistics methods to evaluating mineral resources: block, section and kraging;
- the system runs on PC platform.

The system has been distributed and applied to more than 30 mineral exploration projects all over the country, with much higher productivity than the traditional way.

Computer Aided Editing and Publishing System for Color Geoscience Maps - MAPCAD[12] Has Achieved Remarkable Success In Changing Traditional Working Way

As everybody knows, maps play a very important role in geoscience which are the main representation form of geological concept and achievement, which serve various fields of social development. Their publication, with traditional method, needs about 20 processes, and takes average 1 year or more. It is very difficult to meet demand for on-time map requirement from different fields of national economic development.

Function and features of MAPCAD
MAPCAD is a large-scale application software with copy right of MGMR, which was released in 1993. It makes the whole procedure of map publishing, from digitizing, editing, processing, proofing , color separating until output of plate film , be computerized on PC . The first complex geological map - 1:3500000 geological map of YUN NAN province was printed in Sep. ,1994 as a test map. Having studied and judged carefully, experts from geology, mapping, cartography, map publishing and computer graph got an unanimous conclusion that MAPCAD is powerful enough for color geological map open publishing(in color, accuracy, fineness and etc.).The main features of the system are as follows.
- Graphic data can be input by both digitizer and scanner with effective vectorization function and deviation correcting;

Proc. 30th Int'l Geol. Congr., Vol.25, pp. 145-154
Zhao Peng-Da et al (Ed.)
© VSP 1997

JAFOV Database on the WWW Service

KAICHIRO YAMAMOTO
Koka Women's College, 38 Kadonocho, Nishikyogoku, Ukyo, Kyoto 571, Japan
NIICHI NISHIWAKI
Nara University, 1500 Misasagicho, Nara 631, Japan

Abstract

The JAFOV is the database composing about 4,500 descriptive data on the vertebrate fossil specimens deposited in Japan. It has been constructed since 1982, and stored at Kyoto University Data Processing Center, Kyoto, Japan, as an on-line database using a mainframe computer. It is, however, not easy to use and difficult to serve images in this system. So we tried to apply the WWW technique to improve those problems. The WWW is well known as a very functional way to serve various kinds of multimedia information on the Internet and as providing excellent user interface. In this study, we developed a prototype of new JAFOV service system using WWW technique for serving it on the Internet. Consequently, we confirmed that the new system is suitable for serving such kind of database as JAFOV.

Keywords: database, fossil, vertebrate, specimen, Internet, WWW, DBMS

INTRODUCTION

The JAFOV is the database composing descriptive data on vertebrate fossil specimens deposited in Japan. It has been constructed since 1982, and about 4,500 specimens are recorded in the database [1, 2, 3, 4, 5, 6] . The database is stored at Kyoto University Data Processing Center, Kyoto, Japan, and served as an on-line database using a mainframe computer. The service has, however, some problems such as not easy to use and difficult to serve images (photographs or sketches of fossils) on line.

In this study, we tried to apply WWW techniques to make it easy to use and possible to serve images on line. WWW (World Wide Web) is well known as a very functional way to serve various kinds of multimedia information on Internet and as providing excellent user interfaces. By linking it with DBMS, we can get a much better service system for the JAFOV than the on-line database system currently used.

The followings are the requested specifications of the system developed in this study:
a) The data can be retrieved by using the WWW browser such as Mosaic, Netscape, and others.

b) Not only documentary and numerical data but also image data of fossils can be treated.

c) Retrieved data can be directly displayed on the terminal, printed out, and down loaded to user's computer.

OVERVIEW OF JAFOV

Database Contents

The name JAFOV is derived from JApanese FOssil Vertebrate. It is a database of fossil specimens of vertebrates deposited in Japan. The JAFOV was designed as a database containing document, numeric and image data on fossil specimens as shown in Fig. 1. However, only document data have already been prepared. Other type of data, numeric and image, had been discussed to be included into the database, but it has not yet been realized so far mainly due to technical problems.

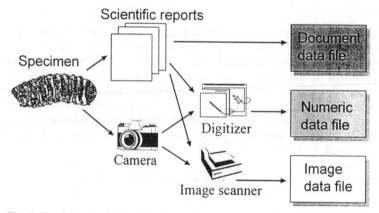

Fig. 1. Total image of JAFOV database designed initially

The JAFOV uses a hierarchical model type of DBMS named FAIRS for mainframe computers manufactured by Fujitsu Co. Ltd. [7]. That DBMS is suitable for documents database such as the JAFOV currently served. It makes an inverted file for items used frequently as search keys to speed up the search process. It is not suitable for the treatment of numerical and image data.

The JAFOV database consists of 41 items (Table 1) including description of specimen itself, geographical location, geological horizon, depository, and others. They can be classified into 8 groups, those are name, classification, locality, horizon yielding the fossil, geologic age, portion of fossil, depository of specimen, and bibliographies describing it. Some of them are defined as search keys, and the others are searched as text items except for a few items only for output.

Most of these items are directly entered from original data, but others can be generated from other items by using dictionaries and/or conversion tables. Some of them are produced referring the tables prepared as shown by arrows in the table. Some are produced by being extracted from the parent items also as shown in the table. They reduce data entry

Table 1. Recorded itmes in JAFOV database

Registration No.	Country	Era
Type	Prefecture	Period
Model	City et al	Epoch
Phylum	Map(1:200000)	Stage
Class	Map(1:50000)	Radiometric age
Order	Map(1:25000)	Magnetostratigraphic age
Family	Map code	Biostratigraphic age
Taxonomic code	Longitude	Remarks of age
Name	Latitude	Portion
Genus	Group	Remarks of portion
Species	Formation	Depository
Author	Member	Address
Year	Bed	References
Synonym	Remarks of geol.	

work and decrease data error considerably.

Database Constraction

The JAFOV database is constructed and maintained by the JAFOV working group which is a voluntary group organized in AVPJ (Association of Vertebrate Paleontologists of Japan).

The procedure of constructing the JAFOV database is shown in Fig. 2. The original data are offered by contributors in paleontological institutes and museums in Japan. They fill data sheets about their specimens and send them to the working group. The working group

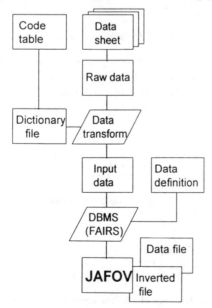

Fig. 2. Procedure of constructing JAFOV databse

examines them and enter the contents into computers to prepare the raw data. The raw data are automatically converted to the input data for the JAFOV by the data transformation program. In this process, some items are produced referring dictionary file obtained from code tables, and some are extracted from their parent items. Then data list is printed out in the JAFOV format, which is sent back to its contributor for data cleaning. If necessary, the raw data are corrected according to the contributor's specification. Consequently, the input data for the JAFOV are prepared. The input data are put into the JAFOV by DBMS according to their data definitions. At that time, two files, data file and its inverted file are constructed in the database.

Current services
The JAFOV is served currently at the Kyoto University Data Processing Center as an on-line database using a mainframe computer. The general image of the services currently carried out is shown in Fig. 3. The database can be used by connecting a terminal to the center either directly or through other computer center. Access to the database is only possible from computer centers connected to the inter-university computer network (NACSIS) before the Internet was constructed a few years ago. Moreover, it needs user identification and is charged to use.

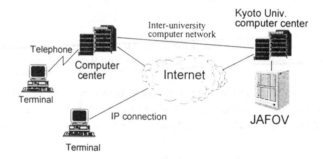

Fig. 3. General image of JAFOV service currently carried out

Fig. 4 shows the usage of the JAFOV currently served. The user connects his terminal to the host computer at the Data Processing Center of Kyoto University by using the telnet directly or through some other computer center mentioned above, logon the computer as a telnet session, then he searches the data he wants interactively by using commands as shown in the figure.

```
LOGON TSS A12345      ← user ID
PASSWORD ########
# IRS JAFOV
RS>SEARCH NAME=*NAUMANNI*
2 SPECIES FOUND
RS>OUT
    ...........
RS>END
FAIRS>END
# LOGOFF
```

Fig. 4. Usage of JAFOV currently served

WWW VERSION OF THE JAFOV

Points to be improved and their solution
The current system of the JAFOV has the following problems especially in its service way, and should be improved to be much more easily and widely used and serve the data other than documents.
a) Poor user interface: The user interface currently used is a command style one (Fig. 4) which is not familiar with any researcher.
b) Restricted service: It needs a registration to any computer center in the NACSIS before using the JAFOV, that is, only the registered user can use the database.
c) Difficulty of constructing and servicing a multimedia database.
d) It costs much to construct and serve the JAFOV on a mainframe computer.

The WWW techniques offer us good solution for these problem as follows:
a) Excellent GUI interface of WWW browser can be used.
b) Binary data transformation is easy through the Internet.
c) Good viewers of image file are provided.
d) Wide access is available by using Internet.

We tried to develop WWW version of JAFOV as its solution. We show its specification and function below.

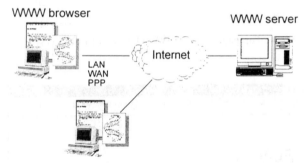

Fig. 5. General image of WWW service of JAFOV

General Image of the WWW service of the JAFOV
The general image of the WWW service of the JAFOV is shown in Fig. 5. When the user accesses to the server through Internet by using the WWW browser, the top page of JAFOV is displayed on his computer (Fig. 6). He can input the search conditions on this page and submit them, then, the records/specimens satisfying the conditions are retrieved and their registration numbers are displayed on the client computer as clickable buttons (Fig. 7). The contents of each record are displayed as shown in Fig. 8, when the button of registration number is clicked.

This example shows the same search procedure as the example in Fig. 4. It may be well recognized that the usage shown here is much more user friendly than the current one.

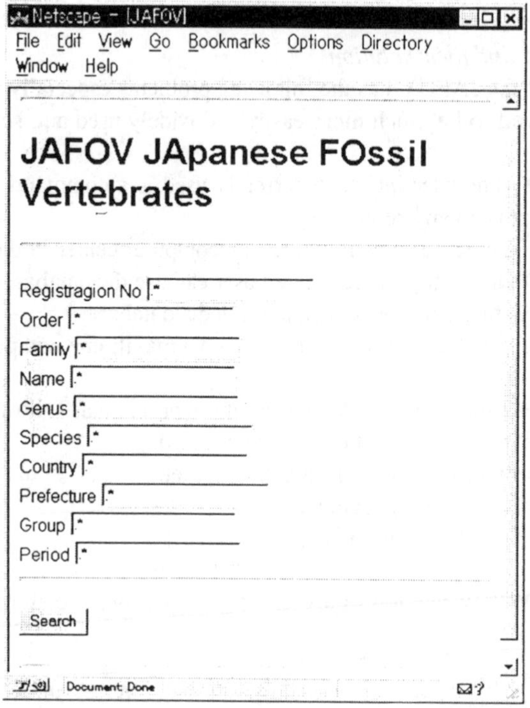

Fig. 6. Top page of the WWW service of the JAFOV

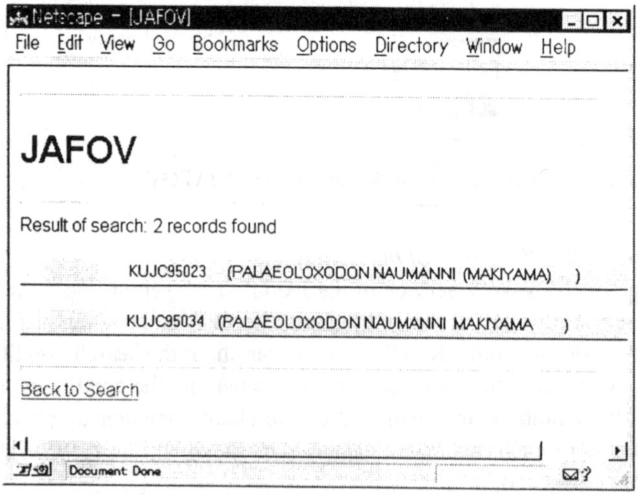

Fig. 7. List of records found as a result of searching the JAFOV

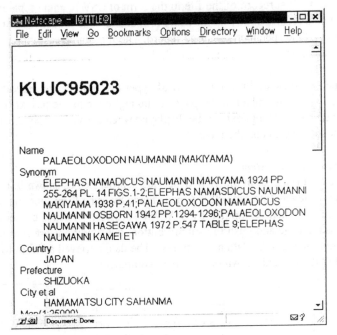

Fig. 8. Contents of record found

DBMS for the WWW version of the JAFOV

In this study, we constructed a prototype database of the WWW version of the JAFOV by developing an original DBMS. The structure of the WWW version of the JAFOV is shown in Fig. 9. It consists of two types of files, that is, the main file and the additional files. The main file contains document data and the file names of image data. Long text data like references can be also stored separately. In those case, their file names are recorded in the main file. Though those type of data can not be searched, it makes search time much short. The image data are prepared as additional files.

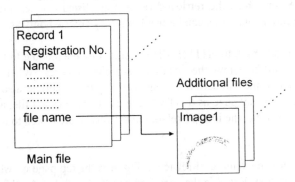

Fig. 9. Structure of the WWW version of JAFOV database

The main file is a simple one of text type in which the items/fields are defined. It consists of the items/fields definition record and specimen data records. The items/fields definition record is put on the first record of the file in the form of CSV (comma separated variable). The specimen data records follow in the same format and same sequence as the items/fields definition record. One specimen uses one record. The file can be easily prepared by using any kind of text editor.

The image and text files are linked as a sort of hyper text in the page generated as a result of retrieve. An anchor is embed in the page as linking to the image or text file. The target data containing image or long text can be displayed when the anchor is clicked. The anchor is shown a clickable button on the page.

Process of search in the system

In general, the data retrieval in the WWW service is processed as shown in Fig. 10 via the CGI (Common Gateway Interface) [8]. A page made by using form function of the HTML is displayed on the client computer by a WWW browser. When the user enter search conditions into the page and push the submit button, the conditions are sent to the DBMS via the CGI interface and the data are searched. The data retrieved are sent to the client as a form of a HTML file, and shown on the client computer.

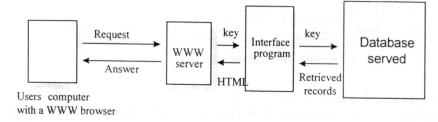

Fig. 10. General mechanism for serving information retrieval by the WWW

The mechanism of searching the database and displaying the results in the system developed is shown in Fig. 11. The system also uses the CGI interface basically. First of all, the conditions entered by the user are sent to the data search module by the CGI. The module searches the data in the main data file, and makes a temporary file for selected data and a HTML file on which the retrieved records are listed as clickable buttons (Fig. 7). More than one search conditions can be used, but they are used only as 'and' condition.

Then the server sends back the HTML file generated by the module to the client. The user can click the record to show the detail of searched records. When he clicks any of registration numbers, the indication is sent to the display module also by the CGI. The module makes the display page as a HTML file using the selected file, and sends the page to the client. At that time, the images related to the record are linked to the page using their file names.

The Figs. 6 to 8 are an example of retrieval. Fig. 6 is the top page on which the user enter the searching condition. Fig. 7 is the searched result obtained under the condition that the item 'Name' contains the word 'NAUMANNI'. Two specimens are found and displayed as clickable buttons as shown in the figure. This page is generated by the data search module in Fig. 11. At the same time, the whole contents of data retrieved are saved in the selected

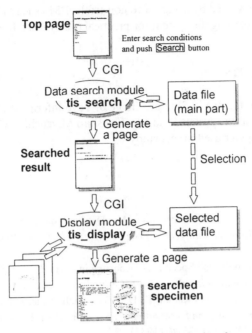

Fig. 11. Mechanism of searching and displaying the records

data file in Fig. 11. Fig. 8 is the contents of record displayed as the result of clicking the upper button in Fig. 7. This page is generated by the display module in Fig. 11 which retrieves the record from the selected file and makes the page by using a template.

CONCLUDING REMARKS

In this study, we developed a prototype system to improve the JAFOV database and its way of services by using the WWW techniques on the Internet in the following points:
a) The user interface is improved considerably, and it becomes much easier to use.
b) The database will be used more widely, for the WWW services do not need any registration in advance and can be used anywhere the client can be connected to the Internet.
c) The database can easily process the image type of data as well as the document type of data.
These improvements are considered to solve the most of problems in the current version of the JAFOV.

There are, however, left some problems to be solved for serving the JAFOV actually by this system on the Internet. The points to be solved are as follows:
a) It takes a little long time to search the data requested.
b) The capacity of the DBMS developed in this study is up to 1,000 records practically. It takes very long time to search the data when the number of stored data exceeds this limit.

These problems will be solved by using more powerful DBMS such as RDBMS as a search engine. So, we are developing the system for practical use by using RDBMS.

ACKNOWLEDGMENTS

We would like to thank Mr. Shintaro Inoue of Toyo Information System Co. Ltd. for his assistance to develop the system, and Mr. Koushiro Miyauchi of CSK Co. Ltd. for his efforts to tune the computer used in this study.

REFERENCES

1. Kamei, T., Yamamoto, K., and Nishiwaki, N., Database on Fossil Vertebrate Specimens Deposited in Japan: JAFOV. *Bull. Data Proc. Cent., Kyoto Univ.*, **19:4**, 260-268 (1986, in Japanese).
2. Nishiwaki, N., Database on Fossil Specimens Deposited in Japan. *Proc. 3rd Intern. Conf. Geosci. Inf.orm.(Adelaide, Australia), Australian Mineral Foundation*, **1**, 62-70 (1986).
3. Nishiwaki, N., Yamamoto, K., and Kamei, T., Data Base on the Japanese Fossil Vertebrates. P. S. Glaeser (Ed.) *Data for Science and Technology, Proc. 8th Intern. CODATA Conf. (Jachranka, Poland)*, North-Holland Pub. Co., 75-80 (1982).
4. Yamamoto, K., Nishiwaki, N., and Kamei, T., JAFOV: Data Base on the Japanese Fossil Vertebrates (1). *Geol. Data Proc.* , **7**, 21-30 (1982, in Japanese).
5. Yamamoto, K., Nishiwaki, N., and Kamei, T., Present Status and Future Extension of JAFOV: Database on the Japanese Fossil Vertebrates. *Geol. Data Proc.*, **12**, 142-150 (1987, in Japanese).
6. Yamamoto, K., Nishiwaki, N., and Kawamura, Y., An Extension of the Japanese Fossil Vertebrates Database JAFOV. *Bull. Data Proc. Center, Kyoto Univ.*, **27:3**, 117-120. (1994, in Japanese)
7. Horiike, H., Ozawa, Y., Murao, Y., and Watanabe, T., *User's Manual: Database Retrieval with FAIRS*. Data Proc. Cent., Kyoto Univ. (1984, in Japanese).
8. Aranson, L, *HTML Manual of Style*. Ziff-Davis Press, Emeryville, California (1994).

Proc. 30th Int'l Geol. Congr., *Vol.25*, pp. 155-162
Zhao Peng-Da *et al* (Ed.)
© VSP 1997

Management Information System (MIS)[1] for The 30th International Geological Congress (IGC), (Aug. 4-14, 1996)

DAI AIDE, LIU XINZHU, ZHOU QI and   HAN MEI

Computer Centre, Chinese Academy of Geological Sciences, 26 Baiwanzhuang Road, Beijing 100037, CHINA

Abstract

The Management Information System for the 30th IGC is one of the scientific research projects supported by the Land Use Department of the State Planning Commission. People's Republic of China. The aim of this research project is to provide a practical management information system for the 30th IGC. In consideration of the reliability. safety. uniformity and timelyness of data, as well as habits of users and economy. a PC local area Network with 10 Base-T and 100 Base-T switch technology is set up constituted by microcomputers of Intel Pentium / 80486 CPU series. MS Windows NT Server 3.51 and Windows for Workgroups 3.11 are adopted for the Network.   The Network management is performed through the Domain Model, so that the centralized safe management of accounts and share in resources will be more structuralized in distribution and use. Main domain and spare-copy domain guarantee immediate restoration in case the data were damaged. MS SQL SERVER 6.0 and MS ACCESS 2.0 is used for the Database Plate running through the Network. In General Code Base and Congress Information Base. the participants' registrations and expenses. academic subjects. abstracts. financial supports. field trips. meeting schedules. cultural activities. etc., are placed on file. Graphical information inquiry interface (GUI) is set for all users.

Keywords: IGC, Computer Network, MIS, Database

PROBLEM BEING RAISED

The 30th IGC held in Beijing (Aug. 4-14, 1996) is an international scientific congress of unprecedented large size ever held in China and hosted by relevant Chinese organizations and governmental departments. More than 9000 participants have registered, and more than 8000 papers submitted, dealing with over 200 various subjects. The Congress covers a vast scope, including more than 50 short courses and workshops, over 70 field study tour itineraries, as well as scientific exhibits, continuous and repeated films and video tapes run in the science theatre, and activities for accompanying young people [28-30]. The Secretariat Bureau of the Congress and the seven special committees are faced with a great quantity of data processing and

---

[1]   Project financially supported by the Land Use Department of the State Planning Commission. China: Study and Development for the Management Information System of the 30th IGC.

information management under their responsibilities of organizing the above-mentioned activities. In this case, computerization of the complicated and interrelated organizing and management services in inevitable.

The MIS consists of two parts: the main database and its subordinate network of local-area databases. The point lies in how to set up the system and put it into operation. Practically the System is an assemblage of hardware, software, network and database techniques in multipurpose application [1-2]. With the rapid development of computer technology in recent years, large conferences in various fields have adopted computerized system in the work of organization and management. However, 1) there have not yet been found any such kind of software which reach the commercialized level. If commercialized ones abroad can be used on lease, the hardware platform to fit should be put under consideration, at the same time a training course is to be open for learning the better use and maintenance of the system. It is quite expensive and the function of the system may not be guaranteed to be adaptable to the special need. 2) Since computer technology has been developed very rapidly, the hardware platform is renewed nearly every two years. The management system software already in use should be also renewed in order to fit the new hardware platform. And 3) the early-stage system is quite unreliable in safety and mostly has no graphic interface for users. Therefore, the problem is raised that it is necessary to design and develop a new special set of congress management system for the Congress, based on the 1990s' technology of computer network and database.

BASIC REQUIREMENTS

The working staff of the Congress require of a computerized management information system for handling organizing matters and quick information inquiries as follows: 1). The Bureau-level LAN for the Organizing Committee (Precongress). A network at the level of Secretariat Bureau of the Organizing Committee, with the registration and information centre (i.e., the main database of the Congress), interconnected with the respective computer system of each office, processing and sharing the data. 2). The on-site computer network of the Congress (during the congress). On-site area of the Congress - the registration area (check-in or on-the-spot registration, payment, alteration, and inquiries), interconnected with the respective computer system of each office area, processing and sharing the data [4].

The MIS is used in the following aspects: Registration processing, Payment processing, Geohost grant program, Scientific program, Field trip program, Operation program, Exhibition program, Social activities, Query [28-30].

OVERALL TECHNOLOGICAL CHALLENGES AND SCHEME
There are several technological challenges in building the MIS. Those technological challenges include:
- Reliability, safety and fault-tolerant ability of the system [1,5].

- Recovery ability and data compatibility of the database.
- Response speed of processing data in real time.

The information system should be designed in an internationally advanced level, with a high reliability and augment ability and without any risk at all. The system should be convenient in operation, with all the functions of routine processings that the users demand. The scheme put into practice is shown in Fig. 1.

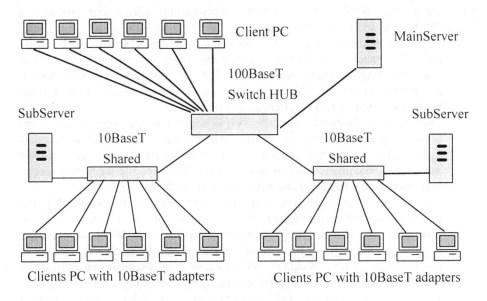

Fig. 1. Topologized schematic diagram of the LAN

Reliability and safety of the LAN hardware

There are many types about LAN technology in the world. One of the best LAN scheme is star type EtherNet for 30th IGC. The 10Base-T and 100Base-T switch Ethernet is selected. The LAN is based on the Intel Pentium / 80486 series PC. As a reliable, safe and fault-tolerant ability of the system, its hardware environment includes 4 servers: 2 main servers (with a memory capacity of 32 Mb) and subservers (with a memory capacity of 32 Mb); 50 client workstations (memory 8 - 16 Mb) ; a number of output devices (HP Laser Printers, or DeskJet Printers); communication equipments (Modem), and other network facilities, which can be augmented and expanded. The LAN topologized structure is an assemblage of 100 BaseT rapid switching HUB and several 10 BaseT shared HUB (Fig. 1). The LAN may obtain a relatively high frequency bandwidth performance (100 Mbps). In addition, it can be integrated tightly with the 10 BaseT Ether LAN used in the early preparatory stage of the Congress, so that the early-stage investments of the users are protected. The high reliable UPS is used for the LAN [4,5].

Reliability and safety of the LAN software
In consideration of the safety, reliability and real time operation of the MIS, the Microsoft network software is mainly adopted for the development and design of the platform. the Client / Server pattern is used for that of the System [6-9, 17,18].

At the terminal of the main server, Microsoft Windows NT 3.51 Server is used. Windows NT came out in 1993 as a 32-bit multipurpose network operation system platform of the Microsoft Inc., and the improved edition Windows NT 3.51 Server was publicized in 1995. Windows NT 3.51 Server is also a fault-tolerant operation system [20,21,26]. In order to intensity the safetyness of the Congress MIS,

1)   The server hardware are specially allocated with double - SCSI Fix-disk (2Gb), meanwhile magnetic-disk managing instruments of Windows NT are used to realize the RAID1-Level processing [20,21,26]. In this way, the fix-disk mirror-image function exhibits a better I/O performance and superiority in mirrorimage towards the guide area and system area of the MIS.

2)   The Domain Mode of Windows NT is adopted in the management of the LAN. The domain server and backup servers are set up. The backup ones may dynamically operate the domain account number database of each backup servers. If something is wrong with the domain server, one of the backup servers may immediately be turned into a domain server [9,20,21,26].

3)   The network safety management function set within the Windows NT (user identification and authentication, the right limits of file visitation, table-of-contents visitation, user's registration time and place, etc.) [20] is put into effect to guarantee the safety of the System's operation and the safety management of the Congress' LAN. Windows NT prossesses an excellent internal safety mechanism and inner-checking design, mainly represented by the following:

•   User identification and authentication. The user account number management system provides with the user's registration and identification. It controls when a certain client may use the computer and which specific one he or she may use, as well as the right limits for clients to use the shared certain resources;

•   Discretionary access control. The NTFS file system performs safety control of catalogues and files. This decreases effectively the possibility of virus invasion because of the control; and

•   Accountability. The errors of the system can be corrected and the wrong operation can be renewed. The System also has the function of monitoring illegal attempts to register.

Netware environment is adopted for the subserver of the registration area [11,12], since some users are accustomed to the Novell Netware. Visits between different networks can be carried out through the NW GATEWAY that the NT provides [26]. The Microsoft Windows for Workgroups 3.11 selected in the first place for the Windows NT client terminals, which is recommended by Microsoft, is mainly used for the workstation of network's client terminal [6]. It is also applicable to the Netware network.

The System makes use of multiple network agreement of such as NWLink, NBF, etc., of the Windows NT. The NWLink (Netware Link) is an agreement that it is compatible with IPX/SPX of the Novell, while the NBF (NetBIOS Frame) is an agreement that in the Windows NT, the NetBIOS may be applied to expand the user interface. Therefore, the expandability of the System increases greatly.

Network management mode
The development of information processing technology increases the dependence of mankind on such technology, which reflects in network techniques as the requirement for centralized management of computer network resources. In this case, the mode of "domain" in Windows NT [20,26] is used in the MIS of the Congress, so that the computer network are put under a centralized management. The domain and the workgroup are two management patterns that should be mentioned in the same breath. The workgroup is a pear-to-pear network pattern, in which computers are in an equal position with one another and play an equal role, though they have their own specified resources and management tasks. Each computer is responsible to keep itself in safety and protect its users' accounts. This brings some inconvenience to the network management. Nevertheless the domain of the Windows NT is a fundamental network management unit, only a logic concept, which has no relationship with the physical connection between computers within the network. One domain consists of one or several servers and client computers, the domain puts both the user-account-number database and safety strategy database under its unified control. The System has adopted a multi-domain model for management.

Recovery ability and data compatibility of the database
The client / server network database model in used for the design. The database development platform in a Microsoft SQL Server for Windows NT 6.0 [19,27]. It is a high-efficiency client/server relationship database management system, giving support to high capacity of routine management, having the capability of keeping the completeness and safety of the data, complication control, routine management, automatic database recovery and convenient data visit. Through the standard ODBC interface, the client terminal may visit the SQL Server using various kinds of software, for example, the Microsoft ACCESS, Excel or Foxpro [13-16, 22-25]. In the system, the client interfaces were developed   by Microsoft ACCESS 2.0 .

Chinese language environment
The CSTAR 2.0 Plus developed by the New Technology Company, Funder Group, is used for the Chinese language environment of the System. It has an excellent performance in Chinese character processing and a satisfactory compatibility with the English operation system and the software in use [25]. Not any other special Chinese-language software is needed.

DATABASE
The database of the Congress is composed of two parts: the client-terminal data utilization software at the front stage and the database server at the backstage (Fig. 2.).

In the back-stage database, the data part includes the original database, the commonly used data code base and the SQL applied program. Among them, the original database is used for memorizing the users' original data, totaling 15 forms, such as registration form, abstract form, financial-aid application form, etc. The commonly used data code base provides with the universal codes of commonly used data, totaling 10 tables, such as those of country / region code, item code for congress activities, code for types of participants, hotel code, etc. [4]. The applied program is used for the management of database, data input, data complication, data retrieval regulation, and data calculation. At the front stage end, various devices the users are familiar with may be used to carry out data operation and retrieval. The Congress System has applied the Microsoft: ACCESS 2.0, to set up the interface for special use by the users [3,10,22-24].

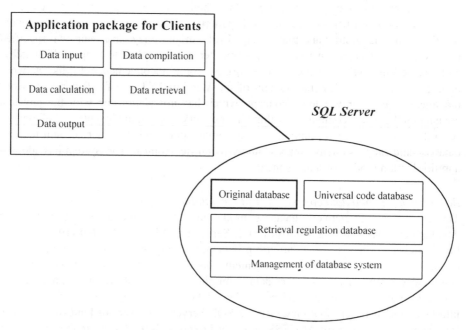

Fig. 2. The overall scheme of the Congress database.

CONCLUSION

Since the MIS of the Congress has been put into operation in April 1994, the System has run in a normal state. The reliability and safety of the System has been raised greatly because of the application of the most recent advanced computer technology of the Nineties. The System has a considerably satisfactory self-guard function, so as to have prevented illegal use of and virus invasion into the main database. When something happens to the System, it can be rapidly switched on and retrieved to keep the completeness and consistency of the data. The client terminal is developed by use of graphic interface and the operation procedure is simplified. All these make the use and operation more convenient for the Organizing Committee of the Congress.

REFERENCES

1. Hu Daoyuan. Computer's Local Area Network, Tsinghua University Publishing House, Beijing. (1990).
2. Wang Yonglin. The System   analysis and Design, Tsinghua University Publishing House, Beijing. (1991).
3. Li Wujun. The Design Technology and Examples of The Database User-Face, Ocean Publishing House, Beijing. (1992).
4. Dai Aide, etc. The Project Program for 30th IGC MIS, (1993).
5. HuDaoyuan. The Guide for The Computer Network, Electric Industry Publishing House, Beijing. (1993).
6. Microsoft Corporation. User' Guide for Microsoft Windows & MS-DOS 6.2, Microsoft Corporation. (1993).
7. Yuan Quanchao. How to Use The Microsoft Project, Electric Industry Publishing House, Beijing. (1993).
8. Hu Yueming. Windows NT Guide, Xueyuan Publishing House, Beijing. (1993).
9. Lu Shiwen. Learn Guide for Microsoft LAN, Xueyuan Publishing House, Beijing. (1993).
10. Liu Qiqqyong. Running Microsoft Access 1.0, Scientific Publishing House, Beijing. (1993).
11. Frank J. Derfler, JR. and Les Freed. How Networks Work, Ziff-Davis Press. (1993).
12. Gu Hong,. The Planning and Design for Novell Network Engineering, Scientific Publishing House, Beijing. (1993).
13. Qin Yilie , 1993. Running Microsoft Foxpro for Windows, Xueyuan Publishing House, Beijing. (1993).
14. Fu Gang, 1993. Programming Design for FoxPro 2.X, Xueyuan Publishing House, Beijing. (1993).
15. Ou Haide, 1993. Development Guide for FoxPro 2.5 for Windows , Xueyuan Publishing House, Beijing. (1993).
16. Zhang Limin. The Analysis for FoxPro 2.5 Command and Function Exampls, Tsinghua University Publishing House, Beijing. (1993).
17. Jim Groves. Windows NT Answer Book, Microsoft PRESS. (1993).
18. Sun Microsystem Computer Corp.. Scope Option And Management Information System. (1993).
19. SYBASE Inc.. The Development Technology for SYBASEClient/Server System. (1993).
20. Cao Kang. The Technology Support for Microsoft Windows NT Server 3.5 , Xueyuan Publishing House, Beijing. (1994).
21. Chen Henan. The Technology for Microsoft NT 3.5, Xueyuan Publishing House, Beijing.   (1994).
22. Microsoft Corporation. User's Guide of Microsoft Access Relational Database Management System for Windows Version 2.0, Microsoft Corporation.   (1994).
23. Microsoft Corporation. Building Applications of Microsoft Access Relational Database Management System
    for Windows Version 2.0, Microsoft Corporation.   (1994).
24. John L. Viescas. Running Microsoft Access 2 for Windows. Microsoft PRESS.   (1994).
25. The New Technology Company of The Funder Group. The User Guide for Chinese Star Ver.2.0 Plus. (1995).
26. Microsoft Corporation. Microsoft Windows NT 3.51 User Guide Book. (1995).
27. Microsoft Corporation. Microsoft SQL Server for Windows NT 6.0 User Guide Book. (1995).
28. The Secretariat Bureau of the 30th IGC. The First Circular for 30th International Geological Congress. (1993).
29. The Secretariat Bureau of the 30th IGC. The Second Circular for 30th International Geological Congress. (1994).
30. The Secretariat Bureau of the 30th IGC. The Third Circular for 30th International Geological Congress. (1995).